An Infinity of Worlds

An Infinity of Worlds

Cosmic Inflation and the Beginning of
the Universe

Will Kinney

The MIT Press

Cambridge, Massachusetts | London, England

The MIT Press would like to thank the anonymous peer reviewers who provided comments on drafts of this book. The generous work of academic experts is essential for establishing the authority and quality of our publications. We acknowledge with gratitude the contributions of these otherwise uncredited readers.

This book was set in ITC Stone Serif Std and ITC Stone Sans Std by New Best-set Typesetters Ltd. Printed and bound in the United States of America.

Library of Congress Cataloging-in-Publication Data

Names: Kinney, Will, author.
Title: An infinity of worlds : cosmic inflation and the beginning of the universe / Will Kinney.
Description: Cambridge, Massachusetts : The MIT Press, [2022] | Includes bibliographical references and index.
Identifiers: LCCN 2021008639 | ISBN 9780262046480 (hardcover)
Subjects: LCSH: Multiverse. | Cosmology—Mathematical models.
Classification: LCC QB981 .K56 2022 | DDC 523.1/2—dc23
LC record available at https://lccn.loc.gov/2021008639

10 9 8 7 6 5 4 3 2 1

For Selene. You are my moon and my stars.

Contents

Preface

There is a story, perhaps apocryphal, about the physicist Neils Bohr and famous Heisenberg uncertainty principle of quantum mechanics. Bohr was known to assert in public lectures that the complementarity under the uncertainty relation between position and momentum in quantum physics—increasing precision in one results in more uncertain knowledge of the other—applies to *everything*. Everything in the world, according to Bohr, has a quantum complement. As the story goes, a member of the audience at one of his lectures asked him afterward, "Professor Bohr, if everything has a complement, then what is the complement of *truth*?" Without missing a beat, Bohr replied, "Clarity." So it is with science writing. There is an inevitable trade-off between making something simple and understandable, and making it complete and fully correct. I hope I have struck an adequate balance in what follows, but I have struck a balance. (I have, for example, chosen to give short shrift to modified theories of gravity, string gas and related cosmological models with thermal boundary conditions, the youngness problem in eternal inflation, and various interesting attempts to define a useful measure of probability on the inflationary multiverse.) For the reader disappointed over a favorite idea glossed over, I can only appeal to the balance between clarity and truth, which is a narrow path I have tried my best to tread honestly. This is not a history of science book, although I have made every effort to include the relevant history of the field as accurately as possible. Likewise, it is not a book dedicated to the philosophy of science, and I attempt to hew as closely as possible to physicist Richard Feynman's belief that "the philosophy of science is as useful to scientists as ornithology is to birds." In this, I ultimately fail; given the nature of the subject, philosophy becomes inescapable.

When I was a young man growing up in the American West, sources of information were few compared to the present day. I had access to only one television station (two if the weather was right), and could not receive Carl Sagan's *Cosmos*, which aired on PBS. My wonderful science and math teachers at Whitefish High School were men who worked as smoke jumpers and rangers in Glacier National Park in their summers off, and their expertise ran more to forestry than astronomy. *Scientific American* was a lifeline for me. The magazine then was written by the researchers themselves, and beautifully filled a gap between journal publications and more popular science. I read it avidly. I have—consciously and unconsciously—borrowed from that early influence here and studiously ignored most of the rules set down by self-appointed experts in "science communication." Instead, I have attempted to emulate the science writing that has had the largest influence on me—hard reads like Steven Pinker's *The Blank Slate*, Richard Dawkins's magnificent pair of books *The Selfish Gene* and *The Extended Phenotype*, Douglas Hofstadter's *Gödel, Escher, Bach*, and Steven Weinberg's *The First Three Minutes*— and have taken stylistic inspiration from writers such as Joan Didion, Annie Dillard, Norman Maclean, and Antoine de Saint-Exupéry. I manage at best a pale echo of these giants. But in the end, the voice is my own, and a part of me is still back in the deep woods of Montana, up late at night with my friends, marveling at the sky alight with luminescent curtains of aurora borealis, popping and crackling above the treetops.

1
The Beginning of the World

> Once upon a time, before ever this world was made,
> there was neither earth nor sea, nor air, nor light,
> but only a great yawning gulf, full of twilight,
> where these things should be.
>
> —Norse creation myth

The universe had a beginning.

Everything we see, everything we ever will see, and everything we are came from this beginning. It began in unimaginably hot fire almost fourteen billion years ago, in which the first elements were forged. The fire cooled, the universe grew dark, and then the first stars flickered into brief lives and filled the cosmos with light. Modern precision cosmology has illuminated the details of our cosmic origins, such as the expansion of the universe, formation of cosmological structure, abundances of the elements, and existence of exotic dark matter, dark energy, and the cosmic microwave background (CMB)—the faint glow left over from the primordial fire. The result is a remarkably consistent picture of the origin and structure of the cosmos known as the *standard cosmological model*, with parameters such as the curvature of space and density of matter measured at the level of a few percent. This physical theory of the hot infant universe, known as the *big bang*, is one of the most consequential developments of science in the twentieth century. But as successful as the big bang picture of the universe has been, it leaves many questions unanswered. Why is the universe so big? Why is it so old? What is the origin of structure in the cosmos? Why was the early universe so simple?

Where did the big bang come from anyway?

The theory of cosmological inflation is our first attempt to answer these existential questions of the cosmos. Inflation can be most simply described as a theory for what happened *before* the primordial fire of the early universe. It is a remarkable new unification of inner space and outer space, in which the physics of the very large (the cosmos) meets the physics of the very small (elementary particles and fields), closing in a full circle at the first moment of time. This new picture of cosmic origins contains quantum uncertainty as a fundamental feature, opening the possibility that the origin of the universe itself was of a quantum nature. In this way, inflation ties questions of the origin of the universe to the unification of gravity and quantum physics, and in so doing, challenges our notions of what we mean by a scientific theory at all. For this reason, the idea has attracted no small amount of controversy. Inflation *is* predictive; the theory has already passed several non-trivial observational tests, and further tests remain outstanding. Nonetheless, the theory of inflation remains a highly speculative idea with a number of difficult to understand implications, such as the possible existence of an eternally self-reproducing multiverse.

In this book, I introduce cosmological inflation as a transformative idea in cosmology that explains previously unexplained features of the universe. In so doing, inflation changes our picture of the basic structure of the cosmos and introduces a set of new, intriguing questions. Inflation provides us with the first observational glimpse of the quantum origins of the universe, and in the end, raises unavoidable questions about the nature of scientific knowledge itself.

The Cosmological Principle

Albert Einstein's general theory of relativity explains the force of gravity as an illusion generated by the geometry of space-time. Because gravity is geometry, any story about the history of the universe is a story about its structure in disguise. Our picture of the universe begins, as with many fundamental ideas in physics, with a principle of symmetry. This is as old as recorded cosmology itself; the universe of the ancient Alexandrian astronomer Ptolemy was built on the ideal symmetry of the sphere, with the Earth motionless at its center and the fixed stars at its outer boundary, the *primum mobile* (figure 1.1). The Ptolemaic universe was structured on Aristotle's system of physics, which held that the universe contains a natural frame of reference—the central Earth—on which all material bodies tend to rest. In the Aristotelian view of dynamics, the motion of material bodies is absolute and does not occur without a cause—an external actor. This viewpoint necessarily requires a separation of the laws of nature that govern material bodies from those that govern the heavens, which are eternally in motion. Medieval astronomers accordingly believed the heavens to be composed of an ethereal substance, inherently different from the substance composing the Earth. This may seem unrecognizable from a modern viewpoint informed by Isaac Newton's laws of motion, but it is a reasonable description of reality. Exert force on an object, and it will move, but then will soon come again to rest. Likewise, there is no a priori reason to believe that bodies in the heavens are subject to the same rules as those on Earth because the Earth has a special position in the cosmos. Despite its symmetry, Ptolemy's universe is a hierarchical one, with the Earth at the center, the lowest point of the universe, and the celestial bodies arranged in concentric spheres above. Humanity manifests in the material world and is subject to physical law. The ethereal world above is forever separate, accessible only to our spiritual bodies. The 1921 Nobel Laureate Anatole France wrote

Schema huius præmiffæ diuifionis Sphærarum.

Figure 1.1
The Ptolemaic universe. *Source*: Commons.wikimedia.org/Fastfission/
Public domain.

in *The Garden of Epicurus* of the universe from the medieval view:

> Looking upwards, he beheld the twelve spheres, first that of the elements, comprising air and fire, then the sphere of the Moon, of Mercury, of Venus, which Dante visited on Good Friday of the year 1300, then those of the Sun, of Mars, of Jupiter, and of Saturn, then the incorruptible firmament, wherein the stars hung fixed like so many lamps. Imagination carried his gaze further still, and his mind's eye discerned in a remoter distance the Ninth Heaven, whither the Saints were translated to glory, the primum mobile or crystalline, and finally the Empyrean, abode of the Blessed, to which, after death, two angels robed in white (as he steadfastly hoped) would bear his soul, as it were a little child, washed by baptism and perfumed with the oil of the last sacraments. In those times God had no other children but mankind, and all His creation was administered after a fashion at once puerile and poetical, like the routine of a vast cathedral. Thus conceived, the Universe was so simple that it was fully and adequately represented, with its true shape and proper motion, in sundry great clocks compacted and painted by the craftsmen of the Middle Ages.[1]

God had no other children but mankind: in Ptolemy's cosmos, we were both unique, and lesser, beings, alone in the universe.

Copernicus changed it all. He wrenched the Earth from its place and set it in motion in the cosmos by proposing the *Copernican principle:*

The Earth has no special position in the Universe.

Copernicus's universe maintained the underlying spherical symmetry of Ptolemy (and indeed he was far from the first to propose a heliocentric cosmology), but he was the first to frame his principle in the language of relativity, recognizing that a consistent heliocentric cosmology required the introduction of the concept of relative motion, stating in his foundational work *de Revolutionibus*, "Every observed change of place is caused by a motion of either the observed object or the observer or, of course, by an unequal displacement of each."[2] This idea, as much as heliocentrism itself, was Copernicus's deep insight about the universe: motion is relative, and no observer is privileged. It was a radical change from the absolute motion of Aristotle, and formed a basis for the later, more precise expressions of relativity by Galileo and eventually Einstein. It also set the stage for Newton's great unification of the Earth and heavens in his laws of motion and gravitation. Newton's realization that the force pulling an apple to the ground is the same force keeping the moon in orbit around the Earth as well as the planets in orbit around the sun completes the destruction of Aristotle's hierarchy between the laws governing terrestrial bodies and those governing celestial ones. We have always lived in the heavens.

Our modern cosmological picture takes the Copernican idea one step further. Rooted in Einstein's theory of gravity, the general theory of relativity, the standard cosmological model was proposed independently by Aleksandr Friedmann, A. Georges Lemaître, Howard P.

Robinson, and Arthur Geoffrey Walker.[3] It is based on a simple extension of the Copernican principle:

> No observer has a special position in the universe.

This is known as the *cosmological principle*, a term coined by E. A. Milne.[4] The cosmological principle implies that the universe must be (at least in a statistical sense) *homogeneous*, or the same in every location, and *isotropic*, or the same in every direction. The implications are profound. The first is that the universe must be without a boundary in space, for such a boundary would represent a special location. The lack of boundary strictly limits the possible geometry of the cosmos: the universe must either be a closed surface, such as the surface of a sphere or torus in higher dimension, or be *infinite* in spatial extent. This lack of a boundary in space does not, however, translate to a lack of a boundary in time. If we apply Einstein's general theory of relativity to a homogeneous, isotropic universe, we find that such a universe must inevitably evolve. We also find that the universe must have had a beginning—a boundary in time beyond which neither time nor space exist.

A Boundary in Time

Is it possible to make sense of a cosmic beginning? It is philosophically difficult for most physicists to accept—and many would vehemently disagree with the first sentence of this chapter. (I will return to this question in the final chapters of the book.) Physical law is based on a chain of cause and effect, with evolution in time described by differential equations. A cosmic beginning appears to require a *first cause*, which by its nature exists outside the boundaries of physical law. Making matters worse is the presence of an initial singularity at the boundary, where physical quantities such as the density and temperature of the universe tend to infinity. The initial singularity is not a peculiarity of any particular cosmological model; famous theorems by Stephen Hawking, and Hawking along with Roger Penrose, showed that initial singularities *always* occur in space-times that satisfy certain general physical properties.[5] Far from being forbidden in general relativity, such beginnings from points of infinity are *generic*. It is widely assumed that singularities represent an inconsistency in Einstein's theory, which will be resolved by some more fundamental theory such as quantum gravity. This remains to be seen.

In addition to the fundamental problem of the initial singularity, the basic picture of the big bang universe leaves open other central questions, mentioned at the start of the chapter. Why is the universe so big? Why is it so old? What is the origin of structure in the cosmos? Why was the early universe so simple? These are not idle questions. In chapter 3, we will see that not only does the standard cosmology not answer these questions but it also *cannot*, for reasons fundamental to relativity itself. To make this precise, I will focus on two observed properties of the universe: its near-perfect homogeneity and thermal equilibrium at early times, and the nearly exact geometric flatness of space. Because gravity is an attractive force, inhomogeneities tend to grow with time. The universe today is locally inhomogeneous; stars and galaxies and planets are vastly more dense than the average cosmological density, which is equivalent to the mass of about five hydrogen atoms per cubic meter of space. Averaged over large scales, however, the homogeneity assumed by the cosmological principle is well supported by observation. Not only does the present universe appear to be homogeneous, but the early universe was highly so. A few hundred thousand years after the big bang, when neutral hydrogen and helium atoms first condensed from the primordial plasma, the universe was smooth to within a few parts in a hundred thousand. Similarly, the deviation of space from exact geometric flatness is constrained by observation to be at most a fraction of a percent.[6] Like inhomogeneity, the curvature of space grows under gravity; a universe with a small initial curvature will become more strongly curved over time. A universe that is geometrically flat to within a tenth of a percent today must have been flat to within one part in a trillion when the universe was a few minutes old. Like homogeneity, this precisely flat geometry is unexplained in the standard big bang, which I explore in more detail in chapter 3.

Both of these basic puzzles of the big bang are questions of initial conditions. Why did the universe start out so geometrically flat and so smooth? What set the initial conditions for the hot, thermal equilibrium universe of the big bang? The modern theory of cosmic initial conditions, called *inflation*, is the subject of this book. Inflation was a period of exponentially rapid expansion that is postulated to have taken place *before* the onset of the hot thermal equilibrium state of the early universe. Inflation, like dark energy today, was a period of expansion apparently dominated by vacuum energy, by which we mean the energy of empty space itself. The universe during inflation was cold, nearly at absolute zero, and empty of everything besides the energy of empty space. That energy is estimated to be of order 10^{15} billion electron volts, which is about a factor of a hundred billion greater than energies probed in particle accelerators today. Because of this extraordinarily high energy, the physics responsible for inflation is potentially related to the physics governing the unification of the strong and electroweak forces, called grand unification, and may even shed light on theories of quantum gravity such as string theory. This "physics of nothing," developed in detail in chapter 4, creates a link between the structure of the present cosmos and physics at the highest energies, via the earliest moments at the beginning of space and time.

The inflationary picture of the early universe involves quantum uncertainty as a central element. Quantum "fuzziness" in the vacuum of space drives the generation of tiny fluctuations in the rapidly expanding inflationary space-time—a process closely related to the Hawking radiation of black holes. Quantum fluctuations of the vacuum, stretched to huge size by the rapid expansion, create a background of tiny fluctuations in the *density* of the primordial cosmic soup. These initial seeds collapse in the later universe to form galaxies, planets, and stars. In this way, inflation not only explains the homogeneous universe of the big bang but neatly explains the *inhomogeneities* as well, as discussed in chapter 5. The consequence is that inflation is directly testable via astrophysical observation. In chapter 6, I summarize the existing evidence for inflation as a consistent cosmological theory and what we can learn in the future.

The quantum nature of inflation not only influences the local properties of the cosmos but its global structure too. When we step back and consider the inflationary universe as a whole, we find that inflation suggests that the structure of the cosmos on the largest scales and at the earliest times is entirely quantum in nature. There is not one big bang but rather infinitely many, embedded in a larger quantum multiverse that undergoes exponential self-reproduction—a process called *eternal inflation*. In chapters 7 and 8, I consider eternal inflation and its consequences. We must come to terms with the meaning of a cosmic theory that predicts that our entire observable universe, everything that we can and will ever be able to see, is one of infinitely many, forever hidden from our view.

To tell the story of inflation, we must begin at the beginning, with the standard big bang model of the universe.

2
The Standard Cosmological Model

There is in the Universe neither center nor circumference.

—Giordano Bruno, *On the Infinite Universe and Worlds*, 1584

Our universe on the largest scales is extraordinarily simple. The cosmological principle tells us that there can be no special place in the universe, which means that the universe must have neither a center nor an edge; if there were a center, then an observer at that center would have a special position, and likewise for an observer near the edge. The equations of general relativity tell us that the cosmos must necessarily be dynamical, either expanding or contracting with time. These features, the absence of a center or boundary, and (in the case of our universe) cosmic expansion, define the basic structure and dynamics of the standard cosmological model. I start by looking at expansion in more detail.

Expansion and the Big Bang

Most people are familiar with the idea that the universe is expanding. When we observe objects beyond our immediate cosmic neighborhood, we find that all of them are receding from us, with more distant objects receding more quickly than nearby ones. This expansion was first noted by the Belgian Jesuit priest Lemaî-tre in 1927 and US astronomer Edwin Hubble in 1929.[1] They found that the recession velocity v of distant objects is proportional to the distance d,

$$v = H_0 d.$$

The constant of proportionality H_0 is known as the *Hubble constant*, and modern observations put its value at around 22 km/s for every million light-years of distance. (At the time of this writing, there is considerable controversy about the precise value of H_0, with some measurements giving a value of 20.6 km/s/Mly, and others giving 22.9 km/s/Mly.)[2] Misconceptions about the nature of cosmic expansion are common, though. One is the idea that the big bang was a sort of explosion of a primordial "cosmic egg," with stars and galaxies flying away from the original point of the explosion. This clearly contradicts the cosmological principle since such a center would constitute a special position defined by the outward velocity of expansion. Another is that cosmic expansion means that the universe must be expanding *into* something outside the universe itself—a misconception that is fueled by the commonly used analogy of cosmic expansion as the surface of an inflating balloon or interior of a rising loaf of raisin bread.

A more accurate analogy for the universe we live in is to imagine a flat, stretchable sheet of rubber, with an even grid of squares drawn on the sheet, as shown in figure 2.1. To satisfy the cosmological principle—the universe must have no center and no boundary—the flat sheet of identical squares must extend *infinitely* in all directions; this will form a two-dimensional analogy for our three-dimensional universe. (The generalization to three dimensions of space is straightforward, by imagining an infinite 3-D space sliced into a grid of identical cubes.)

This sheet of squares (or in three dimensions, cubes) is only one of three possible shapes, or *curvatures* for cosmological space consistent with the cosmological principle, depending on the density of material in the universe. This is defined by a *critical density* that corresponds to about 10^{-26} kilograms, or equivalent to about the mass of five hydrogen atoms, per cubic meter of space. If the *average density* of material in the universe is higher than the critical density, space curves back on itself (called positive curvature) and forms the three-dimensional surface of a sphere in four dimensions. In this case, the universe is in fact not infinite but instead has a finite spatial volume. If the average density is smaller than the critical density, the universe has a negative curvature and is spatially infinite. If the average

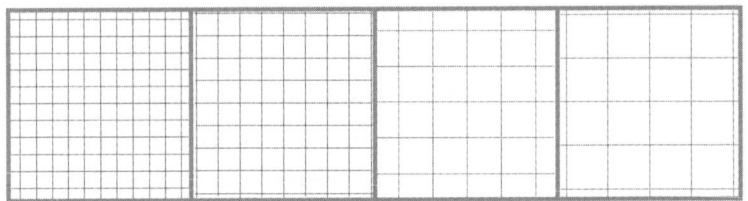

Figure 2.1
Expansion as evenly stretching a flat sheet.

density of material in the universe is *exactly* the critical density, then the universe has a flat geometry, as in figure 2.1. I will discuss later how the curvature of space can be measured, but for now I will simply note that it *has* been measured, and to a remarkable accuracy at that; we know that the overall density of the universe is equal to the critical density to within an uncertainty of about two-tenths of 1 percent.[3] The geometry of the universe is almost exactly flat. Exactly why this is true is something of a mystery, which I will return to later.

Now imagine stretching the flat sheet evenly in all directions so that the squares increase in size, but stay the same shape and all identical, as shown in figure 2.1. The important thing to note is that there is no "center" from which the expansion is taking place; every square is identical to every other square. Galaxies sit (almost) at rest relative to the expanding grid and are swept outward from each other, such that each galaxy sees itself at rest, with all the others moving away uniformly in all directions; *every* observer sees themselves as stationary at the "center" of the expansion. In this way, no location in the universe is privileged, consistent with the cosmological principle. Objects at rest with respect to the expanding grid are called *comoving*. It is not hard to see that such a uniformly expanding grid obeys the Hubble law, such that the recession velocity of a galaxy is proportional to its distance. This is because for a given amount of expansion, the change in the distance between two galaxies is itself proportional to the amount of space between the galaxies. For example, imagine that each square in the grid is at some point in time ten million light-years on a side, and then wait until some later time such that the squares double in size to twenty million light-years on a side. Two galaxies separated by ten million light-years at the start will later be separated by twenty million light-years, but two galaxies separated by twenty million light-years at the start will after the same amount of time be separated by *forty* million light-years. Those initially separated by thirty million light-years will then be separated by sixty million, and so on, such that the rate of change is proportional to the separation at the start. This is the Hubble law.

Cosmological expansion has a dramatic consequence: the universe has a finite age. This can be seen by turning the clock backward, inferring the early history of the universe from its present rate of expansion. Just as our infinite grid of squares grows larger and larger with expansion, if we run the film backward, we see the squares growing smaller and smaller, and if we extrapolate further, the squares will shrink to *zero* size, such that all the material in the universe is squeezed into infinite density. Applying Einstein's general theory of relativity, we find that this approach to zero size and infinite density happens at a finite time in the past. This is called the *initial singularity*, and it is a point at which known physical law breaks down, and our ability to understand or make predictions about the behavior of the universe disappears. Note that the initial singularity is not a point in space but rather a moment in *time*. At this moment in time the universe shrinks to infinite density, but even an infinitesimal moment after the initial singularity the universe is unbounded, and an infinite space springs into being all at once. The big bang happens everywhere at once in an infinite universe!

Another useful way to visualize cosmic history is through a diagram of *space-time*. Einstein's special theory of relativity showed that space and time are really different aspects of the same thing, unifying three dimensions of space and one dimension of time into a single four-dimensional structure. We can sketch space-time by considering two directions of space (suppressing the third) and one direction of time, and drawing it on a plot as in figure 2.2. Observers (you, for example) move along paths through space and time called *world lines*. A single point in space and time is called an *event*. World lines of photons, or anything else traveling at the speed of light, are special. If we plot using the correct units, such as distance in light-years and time in

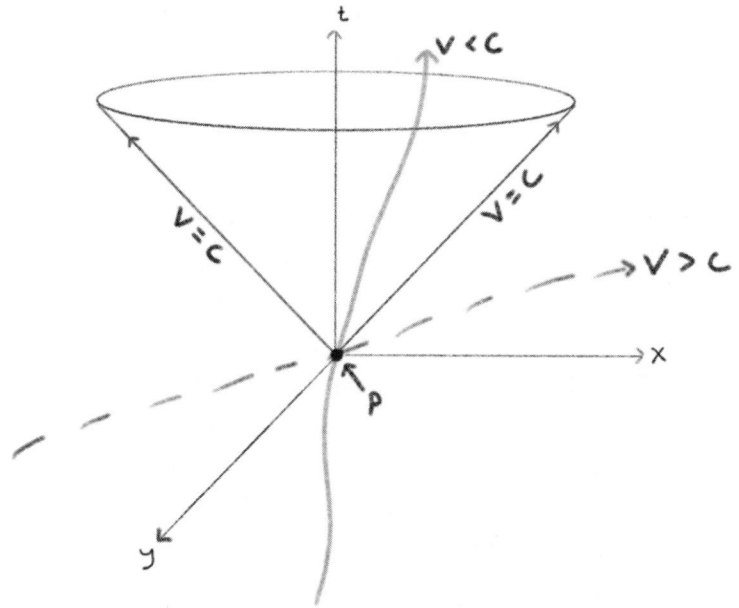

Figure 2.2

World lines in space-time passing through an event labeled P. Light moving at v = c traces out a "light cone" in space-time, while objects moving at v < c remain inside the light cone. Motion outside the light cone from P requires superluminal velocity, v > c.

years, light travels at a speed of one light-year per year, or a forty-five-degree angle on a plot of space and time. Therefore waves from a source traveling outward in circles make a conical shape in a plot of space and time, called the *light cone* (figure 2.2). Objects traveling at the speed of light travel on the light cone, and objects traveling less than the speed of light from the source must remain inside the light cone. To travel *outside* the light cone from a source would require superluminal speed and is forbidden in relativity. This means that a given point in space and time (event) can only send messages to other points in space and time that lie in or on the light cone extending from that event into the future. In this way, the light cone defines the causal future of the event. Similarly, light traveling toward an event in space-time sweeps out a conical surface extending into the past, and the interior of this past light cone defines the set of events that can have a causal effect on an observer at that position in space and time. This *past light cone* defines the causal past of an event. Points in space and time not inside the past or future light cones, called *elsewhere*, cannot have a causal effect on, or be causally affected by, the event (figure 2.3). In the static space-time of special relativity, the future and past light cones extend infinitely forward and backward in time.

We can plot light cones in a cosmological space-time, in which space itself expands with time, by using a trick. As above, we measure spatial distances in units of light-years, but to correct for expansion, we measure time using a clock that slows down with expansion (called *conformal* time) at a rate such that light moving through the expanding space sweeps out forward and past cones at forty-five-degree angles, exactly as in the static space of special relativity. The causal future of an event lies within its future light cone, and the causal past of an event lies within its past light cone. There is an important difference between the static space-time of special relativity and the expanding space-time of cosmology:

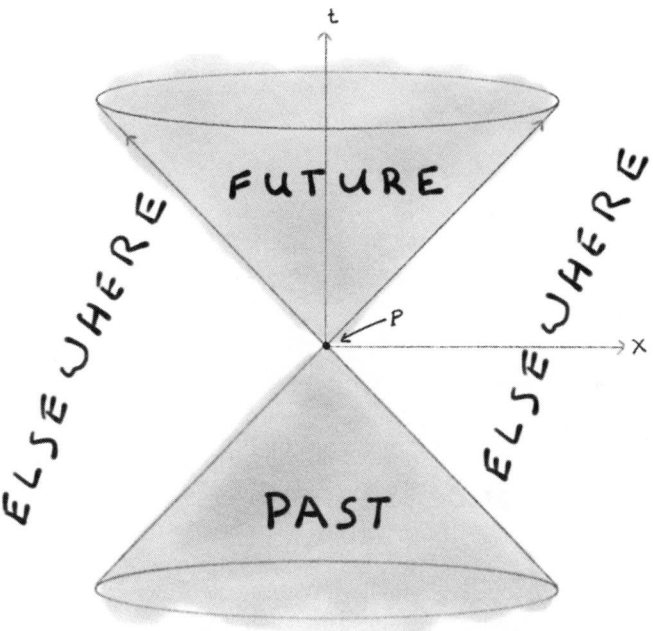

Figure 2.3

The past and future light cones of an event in space-time define its causal past and future. The region labeled "elsewhere" has no causal connection to the event P.

the universe has a finite age. This means that although the future light cone extends forward infinitely in time, the past light cone is finite, "chopped off" by the initial singularity at the big bang. On a diagram of space-time with two directions of space and one of time, the initial singularity is a two-dimensional sheet, infinite in extent: the big bang happens everywhere in an infinitely large space at once (figure 2.4). The big bang takes the

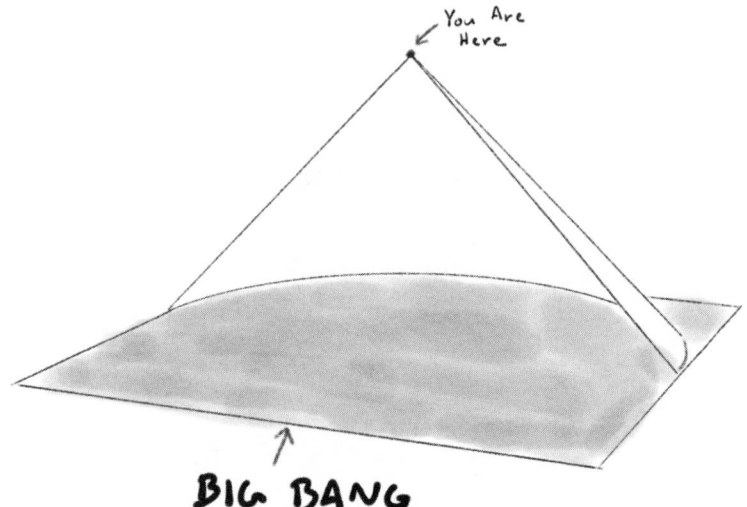

Figure 2.4
The past light cone in a cosmological space-time, which is "cut off" by the spatially infinite initial singularity of the big bang.

infinite, eternal space of René Descartes and hacks off its past with a clean cut, saying "Nothing before this."

Cosmic History: Matter, Vacuum, and Light

With a basic picture of the space expanding forward in time from an initial singularity, we can sketch out a summary of the whole of cosmic history, from the first moment to today, by extending the known laws of physics backward from the present into the past. To complete this description of cosmic expansion, we must also specify the *contents* of the universe. We are made of atoms, as are the Earth and sun and whole of the solar system. Applying the cosmological principle, it follows that the composition of the universe, at least on average, should be the same everywhere. When we look out in space we see galaxies made of gas and dust and stars, all apparently made out of the same stuff as us. This is a modern conception; it was the Copernican principle, made concrete by Newton's unification of celestial and earthly motion that created the modern picture of a universe everywhere composed of matter the same as that as we see on Earth. The only problem is that we now know that this is at least partly wrong; most of the universe *isn't* made of atoms but instead of something else. Things made of atoms, like gas and dust and stars and planets, which cosmologists refer to as *baryonic*, are measured to make up only about 5 percent of the total amount of gravitating material in the universe. The remainder of the cosmic material is "dark," which in the sense it is used by cosmologists means that it (as far as we know) does not interact at all with light, neither emitting nor absorbing it, so that it can only be inferred indirectly by its gravitational effect. More confusing still is that the 95 percent of the universe that is dark comes in two types: dark matter and dark energy. These two types differ by how gravity responds to them, and how they in turn respond to gravity.

Dark matter, aside from not interacting with light, interacts with gravity much like baryonic matter. As the universe expands, it becomes more dilute on average, but the inhomogeneities in dark matter grow with time: regions slightly more dense than their surroundings collapse into even more dense objects. It is this collapse of overdense regions in the universe that is responsible for the growth of structure as well as the formation of galaxies and stars. Cosmic structure formation is mostly driven by the gravitational collapse of dark matter, carrying the small percentage of baryonic material along for the ride. (This is of course only roughly true; feedback from baryons is in fact quite important for the physics of small-scale structures like galaxies.) The difference between dark matter and baryons is that dark matter apparently interacts extremely weakly, if at all, even with itself. This interaction is so weak that efforts to directly detect the stuff have so far all failed, thereby leading some to question whether it exists at all and proposing that in fact it is our understanding of gravity that is deficient. The question remains unresolved.

Dark energy is different. Instead of clumping under the influence of gravity like baryonic matter and (the presumed) dark matter, dark energy remains smooth and in fact may have no inhomogeneities at all. Also unlike dark matter and baryons, dark energy does not become more dilute as the universe expands. Because of this, its effect on cosmic expansion is the opposite of dark matter and baryons. Gravity pulls things together, so the expectation is that the mutually attractive gravitational "pull" of the contents of the universe will act to gradually slow its expansion. It was with this expectation that astronomers in the 1990s set out to measure the cosmic deceleration using supernovas as "standard candles"—objects of known intrinsic brightness—to measure cosmic distances. Instead they found that the expansion of the universe is speeding up. The cause of this cosmic acceleration is entirely unknown, but there is a simple guess that would suffice to explain it. Suppose that vacuum—i.e., empty space itself—carries energy. This is not a new idea; it was first proposed by Einstein as a part of his construction of a static, eternal cosmological solution. (Einstein reportedly later called his introduction of such a *cosmological constant* his "greatest blunder.")[4] Nonzero vacuum energy, or something very much like it, would neatly explain the observed cosmic acceleration as well as a number of otherwise difficult to explain features of the observed universe; in fact a few theorists were arguing for a need for the cosmological constant before the unambiguous detection of cosmic acceleration.[5] The current evidence for the existence of dark energy is overwhelming and comes from multiple independent sources.[6]

The third piece of the "cosmic cocktail" is light, or more generally any kind of massless or nearly massless particle such as photons or neutrinos, which we will refer to generically as *radiation*.[7] These particles, which travel at or close to the speed of light, are by far the most abundant in the universe by number—there are on average more than a billion photons for every baryon—but their contribution to the energy budget of the current universe is negligible, or only about one part in ten thousand. As we will see in the next chapter, most of the light in the universe is not in starlight but rather in an almost completely uniform background that is brightest in microwave wavelengths, which is the residual cosmic "glow" left over from the early hot universe, called the *CMB*, which was first detected in 1965. This leftover (or *relic*) light from the young universe is accompanied by a similar population of relic neutrinos, which (unlike the CMB) has yet to be directly detected. The influence of the relic neutrino background can nonetheless be seen indirectly by its effect on cosmic structure.

The emerging "concordance cosmology" of the present cosmos, divided into roughly 26 percent dark matter by mass, 5 percent baryonic matter, and 69 percent dark energy, with trace amounts by density of photons and neutrinos, is referred to as *lambda cold dark matter* (LCDM), and is a highly precise fit to most existing cosmological data.[8] For the purpose of understanding the very early universe, knowing the composition of the present universe allows us to extrapolate backward to when the universe was young. The key property is cosmic expansion: since the universe is expanding today, by turning the clock backward, objects that are far apart at present must have been closer together in the past. The three elements of the cosmic cocktail—vacuum, matter, and light—behave differently as the universe expands, as shown schematically in figure 2.5. Because of Einstein's famous equivalence between energy and mass, what matters for gravity is not the total mass but rather the total *energy* contained in a region of space. Dark matter and baryons exhibit the simplest behavior. Think of a box full of, for example, hydrogen atoms, sitting at rest inside the box. Now make the box bigger. As the box expands, the gas will dilute, with a density that is inversely proportional to the volume of a box. Dark matter—as far as is known—behaves in exactly the same way. Now consider the same box full of photons— quanta of light. The photons will dilute in number exactly the same way as atoms or dark matter, but there is an additional effect: cosmic redshift. In addition to becoming more dilute as the universe expands, pho- tons (or any other form of radiation) *redshift*, or increase in wavelength, as the universe expands, losing energy in the process. This means that radiation dilutes more quickly with expansion than matter. Dark energy behaves differently still. Since the energy is contained in the empty space itself, dark energy does not dilute in density at all, and the total amount of dark energy in the box *increases* as the box expands. If this sounds like it violates the principle of conservation of energy, that's because it does; energy is conserved only *locally* in general relativity, which means that the sum of all the energy flowing into the box across its boundary must exactly balance the energy flowing *out* of the space sur- rounding the box. This leaves a loophole: you can cre- ate energy out of nothing by creating more empty space

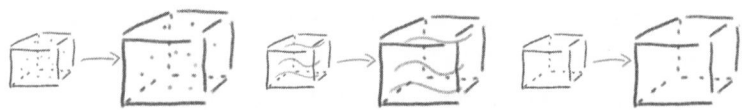

Figure 2.5
Scaling of matter (*left*), radiation (*middle*), and vacuum (*right*) with
cosmological expansion. Matter and radiation both dilute with
expansion, but radiation dilutes faster due to cosmological redshift.
Vacuum, by contrast, does not dilute at all.

inside since nothing flows across the boundary of the
expanding box in the process.

The three types of cosmic energy, with their different
rates of dilution under expansion, determine the basic
framework of cosmic history. The current cosmos is
divided into about three-quarters dark energy, and one-
quarter dark matter and baryonic matter. Only a tiny
fraction of the energy density of the universe today is
in radiation—photons and neutrinos—but these particles are by far the largest contributor by *number*. There
are more than a billion photons in the CMB for every
atom of hydrogen in the universe, and a similar number of neutrinos. If we travel backward in time to the
very early universe, these numbers win; the energy
density of the universe close to the big bang was overwhelmingly dominated by radiation, with matter and
dark energy constituting trace quantities. The density
of radiation dropped rapidly, however, and about sixty
thousand years after the big bang, radiation domination
ended, and dark matter and baryons became the largest contributors to cosmic density, an epoch known as

matter domination. At this time, the scale of the universe was about one-three-thousandth its current size. As the universe continued to expand, the density of matter dropped, until it fell below the floor set by the density of dark energy, which stays constant in time. It was at this point—about eight billion years ago—that the epoch of matter domination ended, and the epoch of dark energy domination began. It was also at this point that the expansion of the universe started to accelerate. Figure 2.6 shows the three epochs of cosmic history in the standard theory of the big bang: radiation domination, matter domination, and finally the epoch of dark energy and cosmic acceleration.

A Cosmic Time Line

The broad cosmic time line of the standard hot big bang discussed in the last section considers the three main epochs of cosmic evolution: radiation domination, matter domination, and the epoch of dark energy and cosmic acceleration. Here I summarize the multiple eras of physics that take place within these three epochs, starting with the era of quantum gravity and ending today. The key figure determining the physics of the early universe is the cosmic temperature; the earlier we go into cosmic history, the higher the temperature, and the higher the energy of the corresponding particle interactions. Thus the physics of the very early universe

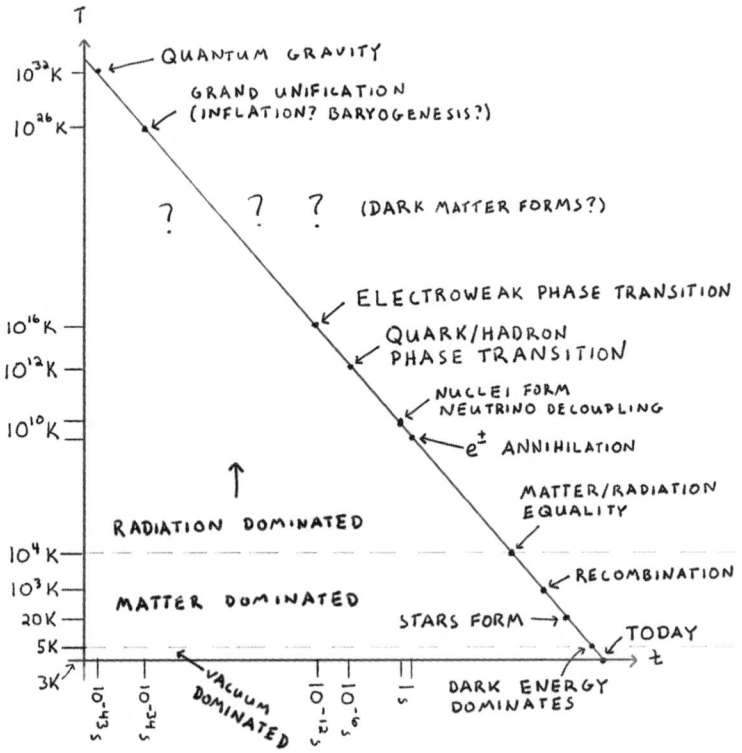

Figure 2.6
A schematic cosmic time line, from the Planck era to today, with temperature on the vertical axis and time on the horizontal.

is the physics of elementary particles and fields, much of which is currently only partially understood. As the expanding universe cooled, unified forces separated in phase transitions, free particles coalesced into bound states, atomic nuclei formed, dark matter, photons, and neutrinos ceased to interact with their environment, and finally, stars formed, galaxies assembled, and the universe evolved into the form we observe today. Here I present a cosmic time line, outlining major events in the early history of the universe.

t < 10^{-43} sec/T > 10^{32} K: The Planck Era

A description of physics earlier than this time requires a theory of quantum gravity, and the physics of this epoch is entirely unknown.

t = 10^{-34} sec/T = 10^{26} K: Grand Unified Theory Symmetry Breaking/Energy Scale of Inflation (Baryogenesis?)

At a cosmic time of around 10^{-34} seconds, the universe is speculated to have gone through a phase transition during which the strong nuclear force separated from the weak and electromagnetic forces, which remained unified until the electroweak phase transition at around 10^{-12} seconds. It is believed that the observed cosmic asymmetry between matter and antimatter—its oppositely charged counterpart—was created at or later than this time, but the mechanism is currently unknown. In theories of inflation, this is the typical temperature of the universe at the *end* of inflation. (In some inflation models, the ending temperature can be much lower.)

10^{-34} sec < t < 10^{-12} sec/10^{26} K > T > 10^{16} K: Unknown (Formation of Dark Matter?)

Between 10^{-34} and 10^{-12} seconds, almost nothing is known about the details of cosmic evolution. It is usually assumed that the universe during this period was radiation dominated, but this is not convincingly established. It is possible that periods of matter domination or inflation occurred—for example, in theories of supersymmetry, where relic particles such as gravitinos are produced in large numbers. It is speculated that dark matter was formed during this period, although it may have been formed well after the electroweak phase transition. During this phase, all the particles in the standard model were massless and behaved as radiation.

$t = 10^{-12}$ sec/T = 10^{16} K: Higgs Instability/Electroweak Phase Transition (Baryogenesis?)

At around 10^{-12} seconds, the Higgs boson became unstable and drove the breaking of the symmetry unifying the electromagnetic and weak forces, and giving standard model particles mass. At this time, the universe was about 10^{-12} seconds old, and consisted of a dense "primordial soup" of free quarks, leptons (electrons and their heavier cousins as well as neutrinos), force carriers (gluons, W and Z bosons, and photons), and their antimatter counterparts, all in thermal equilibrium.[9] Somewhere in this mix was the dark matter, but it is not known for sure at what point, or how, the dark matter condensed from the primordial soup of particles and "decoupled," or ceased to interact. The electroweak phase transition is the highest energy studied so far in the laboratory, at the Large Hadron Collider (LHC) at CERN in Geneva, Switzerland, where protons are collided at an energy of 14 tera electron volts or trillion electron volts (TeV).

$t = 10^{-6}$ sec/T = 10^{12} K: Quark/Hadron Phase Transition

As the universe cooled below 100 million electron volts (MeV), or 10^{12} Kelvin (K), the free quarks and gluons condensed into bound states—protons and neutrons. The universe was about one-millionth of a second old. Uncondensed quark/gluon plasmas have been created in the laboratory at the Relativistic Heavy Ion Collider at Brookhaven National Laboratory in Upton, New York. At this point, the mass inside the observable universe was about equal to one solar mass.

$t = 1$ sec/T = 10^{10} K: Primordial Nucleosynthesis

Below about 1 MeV, a temperature of around ten billion K, the protons and neutrons condensed into atomic nuclei in the epoch of *primordial nucleosynthesis*, when the universe was between about one second and three minutes old, setting the primordial abundances of the elements. In quick succession afterward, neutrinos decoupled from thermal equilibrium with the cosmic plasma, and electrons and positrons annihilated into photons, leaving behind a one part in a billion relic density of electrons. The primordial universe consisted of only the lightest elements: about 75 percent hydrogen, 25 percent helium, and trace quantities of lithium and beryllium. All the heavier elements were formed later in the nuclear fires at the cores of stars, superno-vas, and violent collisions between neutron stars. (Are you wearing a gold ring? Most of the gold was formed in a neutron star collision.)[10] Nucleosynthesis completed at a cosmic time of around three minutes and is the earliest cosmic time that we can say based on observable evidence that the universe must have been radiation dominated.

$t = 60,000$ Years/$T = 10^4$ K: End of Radiation Domination

The universe transitioned from radiation to matter domination at a cosmic age of around sixty thousand years, at which point dark matter, decoupled from the rest of the cosmic plasma, began to gravitationally collapse into bound structures. Baryonic matter (hydrogen and helium nuclei and electrons) remained tightly coupled to photons in the hot primordial plasma and instead set up sound waves called *acoustic oscillations*.

$t = 300,000$ Years/$T = 10^3$ K: Recombination

As the universe cooled below about 3,000 K, the nuclei and free electrons condensed into neutral hydrogen and helium gas. This occurred when the universe was around three hundred thousand years old, liberating the remaining photons into the primordial glow of the CMB (chapter 2), which began to cool and dissipate. This was the onset of the cosmic "dark ages," before the formation of the first stars.

t = 1 Billion Years/T = 20 K: End of the Dark Ages

After the formation of neutral hydrogen and helium gas came the cosmic dark ages, before the formation of the first stars, when the primordial gas gradually cooled and collapsed into the gravitational halos already formed by the dark matter. This cosmic web of dark matter formed the initial skeleton of structure in the universe, with the baryons finally collapsing into the dark matter halos after recombination. The duration of the cosmic dark ages is not precisely known, probably continuing for several hundred million years.[11] Then the first stars lit up, ending the dark ages and beginning the age of the galaxies. The universe was about a billion years old.

t = 6 Billion Years/T = 5 K: End of Matter Domination/ Onset of Cosmic Acceleration

Around eight billion years ago, matter diluted to the point where it was less dense than the cosmic dark energy. Cosmic deceleration then ceased, and the expansion of the universe began to accelerate due to the dark energy. Cosmic structure formation on large scales gradually ceased.

t = 13.8 Billion Years/T = 2.7 K: Today

We are here.

With this cosmic time line, the LCDM cosmology provides us with a consistent, if incomplete, framework for cosmological history, from the big bang to today. At the first moment of time, an infinitely large cosmos emerged from an infinitely dense, infinitely hot singularity. Because the initial state of the universe was extremely energetic, the physics required to understand its behavior is fundamental particles and fields, probed by experiments such as the LHC. Particle colliders like the LHC create conditions that mimic the very early universe. Collisions at the LHC happen at a maximum energy of 14 TeV, which corresponds to a temperature of around 10^{17} K. At higher energies than this—higher temperatures and therefore earlier in cosmic history—the physics is unknown and untested. Our understanding finally breaks down completely at a cosmic age of 10^{-43} seconds, or a temperature of 10^{32} K, when it is believed that space-time itself becomes quantum mechanical, which is called the epoch of *quantum gravity*. Currently, no theory of quantum gravity exists, although several first steps have been proposed—for example, string theory or loop quantum gravity. The universe at this point dissolves into mystery, beyond the limit of human knowledge. There are still many gaps in this long cosmic history, which could be the subject of many books. This book is about the beginning. What set the initial conditions for this cosmic history in the first place?

3
The Cosmic Horizon

> There! the ringed horizon. In that ring Cain struck
> Abel. Sweet work, right work! No? Why then, God,
> mad'st thou the ring?
>
> —Herman Melville, *Moby Dick*

All science begins with what we see. What do we see
when we look out at the universe on a clear night? We
see stars and planets, glowing with reflected light from
the nearest star, the sun. We might conclude, then, that
the universe is made mostly of stars, but this is an illusion, an accident of evolution; we see starlight because
our eyes *evolved* to see starlight, the wavelengths at
which the sun shines brightest. If our eyes were to see
at other wavelengths, we would see a completely different universe: gas and dust shining in the longer
wavelengths of infrared, free electrons spiraling around
galactic magnetic fields in radio waves, or matter falling
into black holes in X-rays and gamma rays. With telescopes, we can sample the full range of available light,

and much of the project of astronomy in the twentieth century was cataloging and mapping the contents of the universe across the full electromagnetic spectrum, extending to the faintest objects deep in the cosmic void. Our picture of the cosmos is increasingly complete. Two main properties of the universe govern the way it appears to us, observing it from the Earth:

1. The speed of light is finite
2. The universe had a beginning and therefore has a finite age

Let us begin with the first of these properties: the speed of light is finite. Because light takes time to reach us from its source, we never "see" the present, but only the past. Watching a baseball player from seats in the outfield, we see the swing before we hear the crack of the bat because light travels more quickly than sound does. We see lightning before we hear the thunder. But although light travels more quickly than sound, it does not travel at infinite speed. The speed of light is frequently expressed in miles per second (186,282) or kilometers per second (299,792), but perhaps a better way to think about it is that light travels 30 centimeters per nanosecond, or one-billionth (0.000000001) of a second, which is about how fast it takes the computer in your cell phone to perform one computation. On ordinary human scales, this is so fast that we can treat light as if it moves instantaneously, yet the scale of the cosmos is so vast that the delay can be immense. When we look at the sun, we see it not as it is now but instead as it *was* eight minutes ago. When we look at the nearest star, we see it as it was four years ago. It is therefore convenient to measure distances in the cosmos by measuring the time it takes light to traverse them; a *light-year* is the distance that light travels in a year, or about 9.5×10^{12} meters. A related, more standard unit of distance in astrophysics, which I use in this book, is the *parsec*, which is a distance of 3.26 light-years. When we look at the center of the galaxy, at a distance of 8 kiloparsecs, we see it as it was about 28,000 years ago. When we look at the nearest large galaxy to ours, the great galaxy Andromeda, a little less than a megaparsec away, we see it as it was 2.5 *million* years ago. Andromeda is the most distant object visible to the naked eye. As of the time of this writing, the most distant observed galaxy was at a distance of almost 13.4 billion light-years, or 4.1 *gigaparsecs* (Gpc).[1] We see it as it was long before the formation of the Earth or even our own galaxy. This is the key fact that makes all cosmology possible:

The further out in space we look, the further back in time we see.

This is profound. The finite speed of light—which is in fact glacial when measured on cosmic scales—means that telescopes function as time machines. We can probe the conditions in the universe one or ten billion years ago because when we look out in space one or ten billion light-years, we can directly see the universe as it was at that time; the light has taken that long to reach us. Unlike archaeologists, who must reconstruct the past from relics in the present, cosmologists can directly observe the entire history of the universe, laid out in sequence from the furthest objects to the nearest. We can literally *see* cosmic history.

The second property that governs the way the universe appears from the Earth is the fact that the universe had a beginning, 13.8 billion years ago. If we could build a telescope that can see out far enough, we could observe the big bang itself! In fact, it is not hard to do, and this—or at least the next best thing—was accomplished in 1965 by Arno Penzias and Robert Wilson at Bell Labs, when they observed the *CMB*, which is the light left over from the early hot remnants of the big bang and has been traveling to us since the universe was a little more than 300,000 years old.[2] If we try to look out further than the age of the universe, what would we see? Nothing, because there is nothing *to* see. Light would have to have been traveling since before the big bang for us to see further. This is the second key fact of cosmology:

The observable universe is finite in every direction.

The "edge" to what we can see is called the cosmic *horizon*, and to visualize this, it is useful to revisit my diagram of cosmic expansion (figure 2.1). Imagine that, just after the instant of the big bang, a beacon emits a flash of light, which travels outward in an ever-expanding sphere, centered on the original point of emission. This is represented in figure 3.1 by a growing circle. At the same time, the space through which the

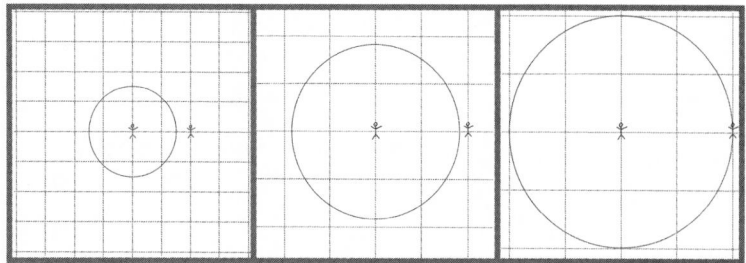

Figure 3.1
The evolution of the cosmic horizon with expansion.

light is traveling expands, as represented by the grid. The distance that the light has traveled since the moment it was emitted is the limit of how far away the light can be seen since the light has yet to arrive for observers in the universe outside the expanding sphere of light. By symmetry, this is also the limit of how far out into the universe observers at the center of the sphere can see; this is the extent of the observable universe as seen from that point. In this example, the expanding sphere of light "catches up" to the expansion: as the universe gets older, our horizon gets larger and larger, and observers at the center see more and more of the universe. Note that—as with the horizon on Earth—the cosmic horizon is an artifact of our viewpoint, not a property of the universe itself. Observers in different locations see different horizons, always with themselves at the center (figure 3.2). The *observable* universe as seen from any point is finite, but the universe as a whole continues outward forever. The universe outside the horizon of an

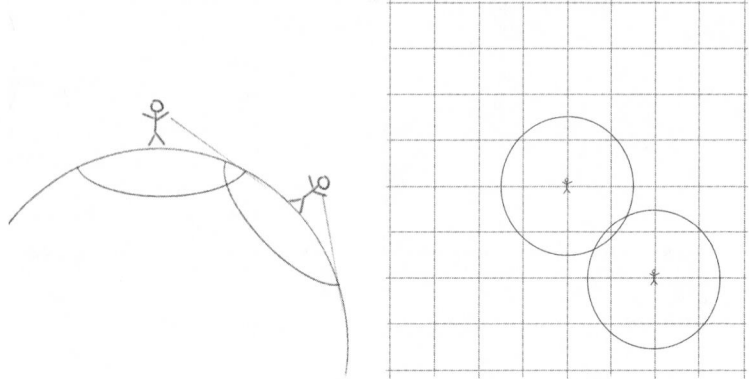

Figure 3.2
The horizon is a function of perception. On the left is the apparent horizon as seen by two observers on Earth. On the right is the apparent horizon as seen by two cosmological observers.

observer at any point is invisible because the light from there has not had time to reach them yet. We can also visualize the cosmic horizon on a space-time diagram of our past light cone (figure 3.3). When we look out in space and back in time, we are seeing light that has traveled to us along our past light cone in space-time. The apparent edge of the universe is then the circle at which our past light cone intersects the surface of the initial singularity at the big bang. Compare this to the Ptolemaic universe in figure 1.1; we are returned to the center of the cosmos, except our firmament is no longer the sphere of the fixed stars but instead the sphere of our apparent horizon and the big bang itself.

It is an interesting fact of gravity that the size of the observable universe is exactly the size of a black hole

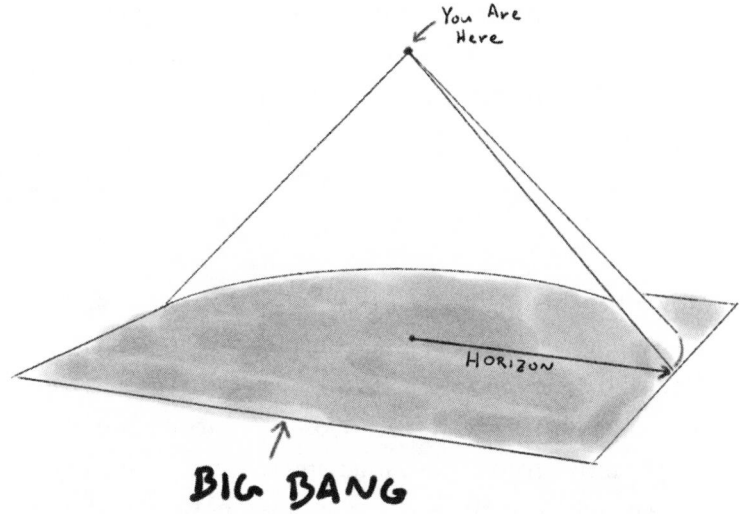

Figure 3.3
The horizon as the projection of our past light cone on the infinite surface of the initial big bang singularity. Light reaching us travels along our past light cone, so the further out in space we see, the further back in time.

with the same total mass, but the universe is *not* a black hole! The horizon of a black hole is a *trapped surface*, for which all physical paths through space-time point inward. The horizon of our universe, dominated by dark energy, is an *antitrapped surface*, for which all physical paths instead point outward, so that things outside our horizon are being swept away from us by expansion faster than the speed of light. This may seem to contradict the core principle of relativity that nothing can travel faster than light. As with the conservation of energy, though, Einstein's general theory of relativity provides a loophole: the distance between two points in space-time may increase at a rate faster than the speed of light if the space itself is expanding.

The Limit of Vision

How does this appear to us, sitting at the center of our own spherical "bubble" of the observable universe? Since looking out in space is looking back in time, as we look out further and further, closer to our horizon, we see the universe as it was earlier and earlier in its history. If we could see all the way to our horizon, we would see the moment of the big bang itself! In practice, however, we cannot see all the way to this ultimate edge of the observable universe because the universe in the early times was very hot and dense, which resulted in the rapid scattering of photons, similar to what happens inside a bank of dense fog. The cosmic medium prior to the formation of stars was astonishingly simple, as mentioned earlier: roughly 75 percent hydrogen, 25 percent helium, and trace amounts of lithium and beryllium. That's it. All the heavier elements were formed later, in the cores of stars, supernovas explosions, or collisions between the compact remnants of supernovas known as neutron stars. The further back in time we go, the hotter the universe gets, and light scattered rapidly in the early universe when this simple primordial gas was above a temperature of about 3,000 K, or about half the temperature of the surface of the sun. Above this temperature, electrons are stripped away from the nuclei of the atoms in a process called *ionization*. This ionized gas, or *plasma*, consisted of positively charged atomic nuclei and highly mobile, negatively charged, free electrons. These free electrons interacted strongly with light, swiftly scattering photons trying to propagate through the plasma and rendering it opaque. The gas itself at this time was still highly rarefied, with a gas density of about 300 million hydrogen atoms per cubic meter, or about one-millionth of a billionth of the density of the comparable plasma at the surface of the sun. The universe at the time would have appeared to the human eye to be a softly glowing, fiercely hot oven. As the universe continued to expand and cool, the free electrons in the hot plasma were captured by the nuclei to form electrically neutral atoms, which are transparent to light, and the "glow" of photons from the hot early plasma began to propagate freely through space (figure 3.4). The time was about 380,000 years after the big bang.

Most of the photons released from the primordial plasma traveled through space undisturbed for the entire 13.8 billion year history of the universe, and as such form a pristine record of the conditions in the primordial plasma. The photons were not, however, *unchanged*; cosmological evolution stretched the photons to longer and longer wavelengths as they propagated through the expanding space. This *cosmological redshift* is frequently misunderstood as being a form of

Doppler shift due to the relative motion of source and detector. The analogy of Doppler shift for cosmological redshift is a good one for relatively nearby objects (where by "nearby" we mean within 100 million light-years or so), but it breaks down at very large distances (billions of light-years), and misrepresents the physical origin of the redshifting of light. Cosmological redshift is a general relativistic effect and due directly to the expansion of space: the wavelength of photons propagating through an expanding space increases proportional to the amount of expansion (the size of the squares in my grid analogy). When the distance between two comoving objects doubles, the wavelength of a freely propagating photon also doubles. This means that photons propagating through the universe form a sort of cosmic "clock": if we know the wavelength of the photon when it was emitted (for example, the wavelength of a line in the spectrum of hydrogen or helium), we can use its observed wavelength to calculate the amount of cosmological expansion and therefore the amount of time it has been traveling to us. This is a powerful tool of observational cosmology and will be integral to many of the results I discuss later in the book.

A second effect of cosmic expansion related to photon redshift is *expansion drag*: since the momentum of a photon is proportional to its frequency, cosmological redshift—a decrease in frequency—can be thought of as a loss of momentum too. The same momentum loss that happens to photons also happens to other particles like atoms and dark matter: cosmological expansion causes freely moving particles to slow down and eventually come to rest relative to the cosmological "rest frame" defined by the expanding grid of squares. In this sense, the expansion of the universe acts as a kind of friction on all moving bodies and explains why most objects in the universe are (almost) at rest relative to the expanding space. This preferred frame of motion in the universe is called the *cosmic rest frame*. Real galaxies in the universe have "peculiar velocities" relative to the cosmic rest frame caused by gravitational interactions with nearby objects, but taken on average, matter in the universe is stationary with respect to the expanding cosmic rest frame. Combined with the presence of an apparent cosmic horizon, this preferred reference frame gives a picture of our universe that remarkably echoes Aristotle's ancient conception of cosmology: we see ourselves at the center of a spherical universe, (almost) stationary with respect to a cosmic frame of rest. Despite this superficial similarity, our modern conception of the cosmos in reality differs from Aristotle's in every detail since it is grounded in relativity—*every* observer in the expanding universe sees exactly the same thing.

When we look out in space and back in time, the oldest light there is to see is the primordial glow of photons released at the moment of recombination, as if we are standing on the ground and looking up at the light filtering through the bottom of a bank of clouds, the space beyond which is obscured from view. We cannot

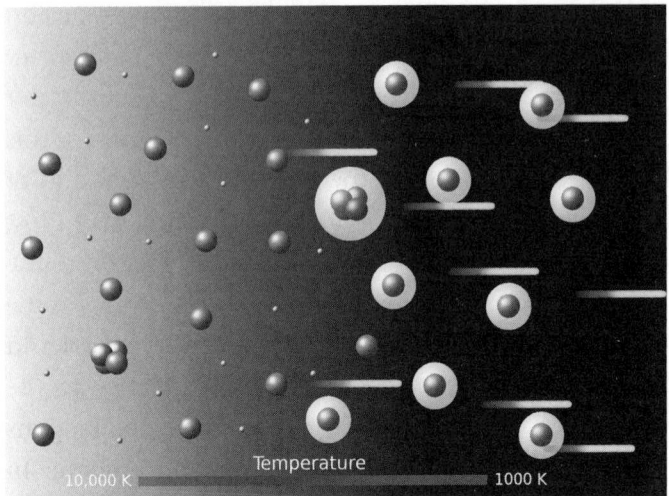

Figure 3.4
A schematic diagram of recombination about 380,000 years after the big bang, when the hot plasma of the early universe (*left*) condensed into neutral atoms and free photons (*right*) at a temperature of around 3,000 K.

see further than this *surface of last scattering* of the photons emitted by the hot primordial gas, in nearly perfect thermal equilibrium at a temperature of 3,000 K. As the universe expands, these photons stretch in wavelength proportional to the expansion, resulting in cooling, and are in the present universe at a temperature 1,100 times lower, or about 2.73 K, which corresponds to microwaves or very high frequency television signals. This CMB arrives on Earth precisely evenly from all directions and appears to us as if the sky is glowing softly in microwaves, and it is this glow that was first observed by Penzias and Wilson at Bell Labs in 1965.[3] (In fact, about 1 percent of the background static at frequencies used for television is photons from the CMB. When you look at the static in a predigital TV, mixed in with the other noise, you are seeing the afterglow from the big bang!)

The surface of last scattering, at least when we are looking at the universe using light, is the edge of our observable universe. It is the limit of vision.

One curious property of the photons arriving from the surface of last scattering is their high degree of uniformity. The CMB is very nearly, but not perfectly, uniform. The largest source of nonuniformity is due to the Earth's motion relative to the cosmic rest frame, caused by our combined motions of the Earth orbiting around the sun, the sun orbiting around the center of the Milky Way galaxy, and the Milky Way itself being pulled through space by the gravitation from the Great Attractor, a huge concentration of dark matter between 150 and 250 million light-years away.[4] This motion relative to the cosmic rest frame causes the photons in the CMB to Doppler shift by a few parts in a thousand, appearing as a slightly shorter wavelength (bluer) in our direction of motion and slightly longer wavelength (redder) away from our direction of motion. This *dipole anisotropy* was first detected by Paul Henry in 1971, and later confirmed by B. E. Corey and David Wilkinson as well as George Smoot, Marc Gorenstein, and R. A. Muller using observations from a U2 spy plane in flights between 1976 and 1978.[5] Our local motion corresponds to a velocity of about 370 kilometers per second in the direction of the constellation Leo and is not an intrinsic property of the primordial radiation from the early universe. If we correct for our local motion, we find that the CMB is *isotropic*, the same in every direction, to within a few parts in 100,000. This has been measured to exquisite accuracy by a large number of experiments and tells us something important about the state of the early universe: if the light emitted by the primordial plasma was extremely uniform, the plasma itself must have been extremely uniform. This is not an inference; we can directly observe it! When we observe the CMB, we are *seeing* the universe when it was a little more than 300,000 years old, and what we see is that the universe was almost perfectly uniform everywhere in space. The property of homogeneity that was assumed as a symmetry to solve Einstein's equations turns out to be extremely well supported by data, which is one piece of compelling evidence for the correctness of the big bang model of the universe. This raises a new question, though. How did the universe get that way?

The Horizon Problem

The high degree of uniformity and thermal equilibrium of the early universe may not seem like a terribly peculiar feature, or even something that needs explaining.

It is only when we consider the evolution of the cosmic horizon in time that it becomes a deep mystery. Remember that throughout most of cosmic history, the size of the observable universe has been growing. Observers watching the universe over the span of billions of years will see more and more space inside their cosmic horizon, but this works in reverse as well: the size of the horizon at early times was *smaller* than it is today. A rough estimate of the size is just the universe's age measured in light-years; when the universe is 1,000 years old, the observable universe is about 1,000 light-years in radius. When the universe is 1 million years old, it's 1 million light-years. This rule of thumb is always a bit of an underestimate because it does not take into account the expansion of space. The universe today is 13.8 billion years old, but the radius of our observable universe is about 40 billion light-years because space has been steadily expanding throughout the almost 14 billion years of cosmic history, and a precise calculation requires taking this expansion into account. In the very early universe, this discrepancy was an exact factor of two, so that when the universe was a year old, the cosmic horizon about any point was 2 light-years in radius.

What does this mean for what we see? When we look at the CMB, we are looking outward a distance of nearly 40 billion light-years and back in time to when the universe first cooled to transparency. When the CMB was emitted, about 300,000 years after the big bang, the observable universe was about 730,000 light-years in radius—a tiny fraction of its size today. The observable universe when the CMB was emitted appears to us today to be a circle on the sky about one degree across, or about twice the size of the full moon. When we look at the CMB sky, we are seeing thousands of causally disconnected horizons—literally separate universes. As long as relativity is correct, there was no possible way for these thousands of separate universes to communicate with each other during the 300,000 years since the big bang given that such communication would have required faster-than-light signals. Yet it is an observational fact that these thousands of separate universes are absolutely identical to each other in density, temperature, composition, and all other physical properties. The variation from one region to the next is a few parts in a 100,000. The standard big bang cosmology offers no explanation for this extraordinary degree of uniformity in the early universe. This is referred to as the *horizon problem* in cosmology.

The horizon problem can also be visualized on a diagram of space-time such as in figure 2.4. When we look out in space and back in time, we are observing further and further back along our own past light cone, and when we look out at the CMB, we are looking nearly all the way back to the initial surface of the big bang. The CMB seen in different directions in the sky are different points (events) on our past light cone. The causal past as seen by observers at those points are the past light cones of those events in space and time. Figure 3.5

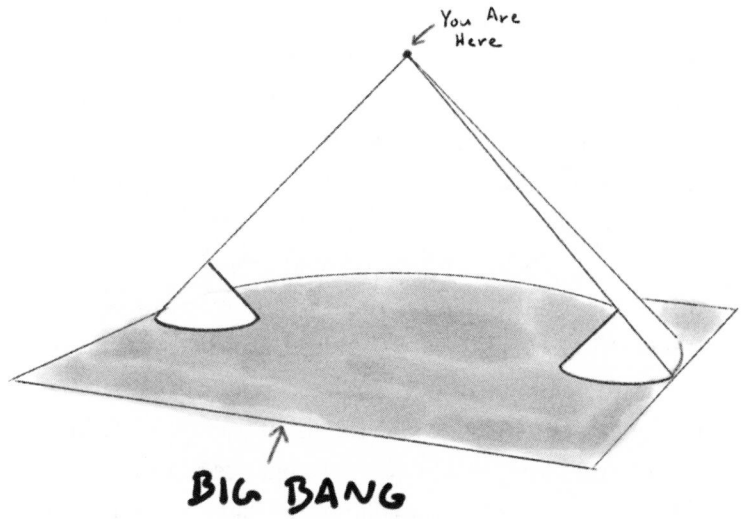

Figure 3.5
Diagram of the horizon problem in space-time. When we observe distant objects, we look out along our past light cone. The past light cones of those points in space and time do not overlap, indicating that they share no causal past connections.

schematically shows the past light cones of two events along our past light cone. If we look far enough back, the past light cones of those distant events do not overlap, indicating that two points on opposite sides of the CMB sky cannot have shared a causal past in the time since the big bang! Our observable universe today consisted of many separate horizons at the time that the CMB was emitted.

There are only a few possible ways to solve the horizon problem. The first and least satisfying is to simply assert without explanation that the initial state of the universe was a highly uniform one. Sometimes seemingly unlikely things do happen by accident. Consider, for example, the fact that the moon and sun as seen from the Earth are almost exactly the same size in the sky—a phenomenon that makes solar eclipses both especially brief and especially spectacular. There is no deep explanation for this coincidence of sizes; it's a fluke. The homogeneity of the universe may be a similar fluke. (In fact, if the early universe were very nonuniform, we would be unlikely to exist at all since an inhomogeneous universe would fragment and collapse into singularities long before life had a chance to form. I will return to such *anthropic* explanations later in the book.) The second possible solution to the horizon problem is that perhaps there is some mechanism for faster-than-light signaling so that the many separate horizons in the early universe were in fact able to exchange signals and reach equilibrium. The third possibility is that something else happened, before the onset of the hot big bang, which set the initial conditions for the subsequent evolution of the universe. It is this third option that forms the subject of this book.

Primordial Anisotropy

What about the remaining anisotropy, the one part in a hundred thousand nonuniformity in the CMB? This was first measured by the Cosmic Background Explorer

(COBE) satellite in 1992, and later with much higher sensitivity and precision by the Wilkinson Microwave Anisotropy Probe (WMAP) and Planck satellites as well as a variety of measurements made from ground-based observatories and high-altitude balloons over a period of nearly three decades.[6] Unlike the one part in a thousand hemispheric anisotropy from the Earth's motion relative to the cosmic rest frame, the remaining anisotropy in the CMB is *primordial*—that is, it is a property of the very early universe. These tiny hot and cold spots—variations in the temperature of the CMB—are caused by variations in the density of the matter in the early universe, particularly regions of high and low density in the dark matter. These *primordial perturbations* are believed to have been present, imprinted in the cosmic dark matter, nearly back to the big bang itself. The primordial perturbations in the dark matter remained dormant until the universe transitioned from radiation to matter domination at an age of about sixty thousand years. At this point, regions that were more dense than their surroundings began to slowly collapse under the force of gravity, pulling dark matter from surrounding regions into "halos," which were the progenitors of all structure in the present universe, such as galaxies and clusters of galaxies. Prior to recombination, the cosmic plasma was too hot to form galaxies because the baryonic matter was still coupled strongly to the radiation and was unable to collapse along with the dark matter halos. The dark matter, because it has little or

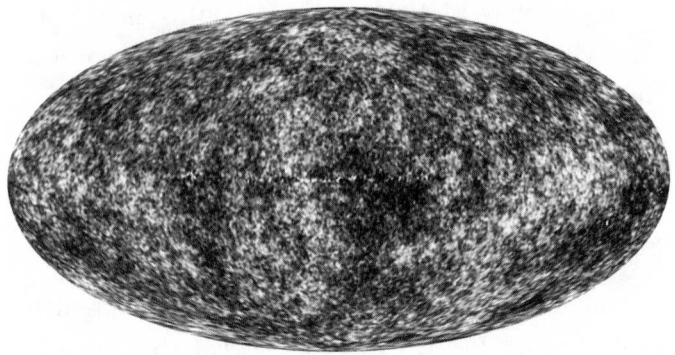

Figure 3.6
The microwave sky as seen by the Wilkinson Microwave Anisotropy Probe (NASA/WMAP Science Team). The light and dark regions correspond to a few parts in a hundred thousand variations in the temperature of the CMB.

no interactions besides gravity, collapses freely, but the baryons interact in two ways: like the dark matter, the baryons tend to collapse under the force of gravity, but interaction between the hot baryonic plasma and photons pushed the baryons apart, and out of the dark matter halos. This seesaw of opposing forces means that the collapsing dark matter halos released *sound waves* in the baryonic plasma, which propagated outward at about one-third the speed of light, and created hot and cold spots in the CMB (figure 3.6). This set up a fascinating piece of physics: the cosmic horizon—then less than a million light-years across—formed a finite cavity for the sound waves, much like the cavity formed by the body of a flute. As in a flute, this cavity created *harmonics*—waves that fit exactly once, twice, or three times into the cosmic horizon. These harmonics, called *baryon acoustic oscillations*, were predicted by Soviet physicist and dissident Andrei Sakharov in 1966, and were first confirmed unequivocally by the TOCO telescope in Chile and later by the BOOMERANG Antarctic balloon experiment.[7]

The harmonics in the primordial sound waves can be seen by plotting the strength of the waves as seen in the CMB at different wavelengths, called a *power spectrum*, as shown in figure 3.7. The peaks in the power spectrum represent the successive wave harmonics, with the line representing a best-fit LCDM cosmology. The agreement is astonishing, and the CMB anisotropy represents powerful evidence in favor of the LCDM cosmological model. In particular, the existence of acoustic oscillations in the CMB is strong evidence for the existence of dark matter since without the presence of dark matter halos, no sound waves would exist, and density perturbations in the baryonic matter would be mostly erased by photon pressure.

Behind the stunning agreement between LCDM cosmology and the CMB lies a fundamental question: Where did the original perturbations in the density of the dark matter come from? The CMB places a tight constraint on the form of these perturbations, particularly that they must be nearly, but not exactly, the same amplitude at every wavelength—a property called *scale*

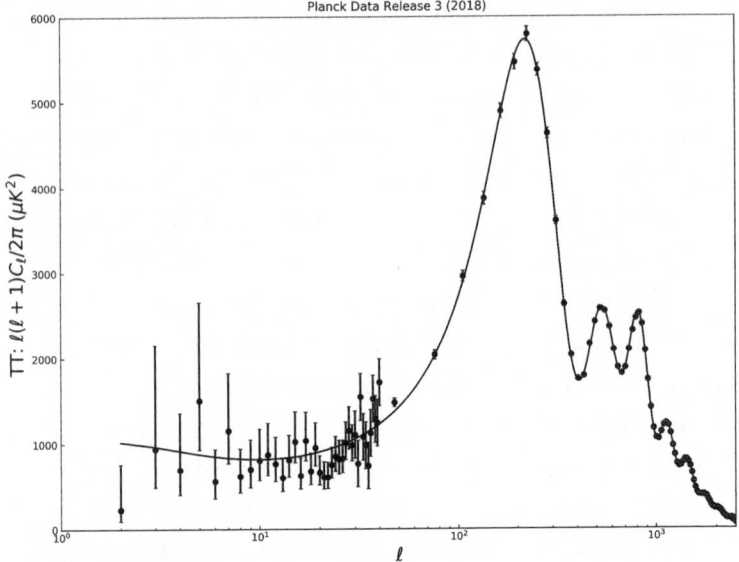

Figure 3.7
A plot showing the astonishingly good fit of the LCDM cosmology to the CMB. The points show the Planck satellite measurement of the CMB anisotropy, plotted as a function of angle on the sky, with vertical bars depicting the measurement uncertainty. Large angles are on the left, and small angles are on the right. The line illustrates the best-fit LCDM cosmology, with multiple peaks showing the presence of sound waves in the primordial plasma. *Sources*: Data: ESA/Planck collaboration; plot by the author.

invariance. The LCDM model assumes approximate scale invariance as an input to the model, but it postulates no mechanism for creating the primordial perturbations in the first place. Just as LCDM contains no explanation for the isotropy of the CMB, it likewise contains no explanation for the *anisotropy.* I will return to this question in chapter 5.

Curvature and the Flatness Problem

The presence of acoustic oscillations in the early universe, itself a remarkable confirmation of the standard cosmological picture, also provides us with an extraordinarily precise tool with which to measure the curvature of space. The best current measurement of the geometry of space comes from the Planck satellite measurement of the CMB anisotropy, with the result that the universe differs from perfectly flat geometry by at most two one-hundredths of a percent—one of the landmark results of modern precision cosmology.[8]

We measure the curvature of space by measuring its geometry. The familiar identity from elementary geometry that the three angles of a triangle add up to 180 degrees is only true on *flat* surfaces. If you draw a large enough triangle on the curved surface of the Earth, you will find that the angles add up to *more* than 180 degrees (in fact, it is possible to draw a "triangle" on the surface of a sphere that consists of three right angles!). The same is true for triangles drawn in a positively curved universe. Conversely, if the universe has a negative curvature, akin to a saddle shape, the angles of a triangle will add up to less than 180 degrees. To measure the curvature of space, we simply need to find a large enough triangle to measure, where the sides of the triangle are of known length, called a "standard ruler." Such a standard ruler is provided by the cosmic horizon at the surface of the last scattering since its radius is determined

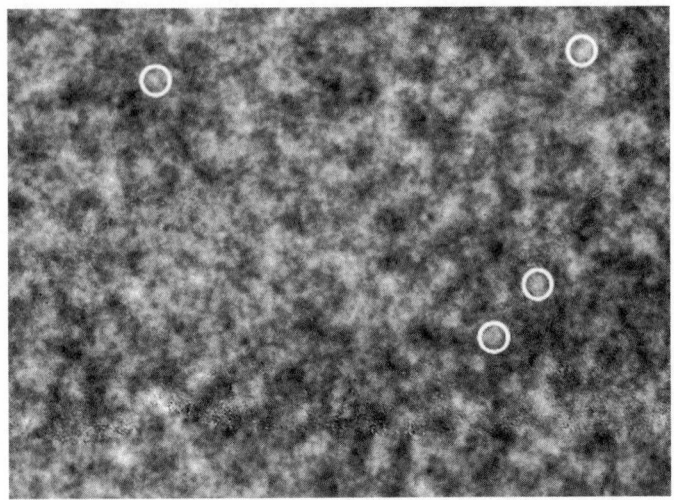

Figure 3.8
A blowup of the Planck map of the anisotropy of the CMB. The hot spots (*bright regions*) have a typical size that corresponds to the horizon size at the time of recombination (*white circles*). These form a standard ruler. *Source*: ESA/Planck collaboration.

by the age of the universe at recombination, which is a calculable quantity. Furthermore, we can detect the presence of this horizon via the tiny anisotropy in the CMB observed by the WMAP and Planck satellites. The fluctuations are random, but they have a characteristic size that is easily visible by the eye (figure 3.8), corresponding to the size of the cosmic horizon at that time. (More precisely, this corresponds to the size of the *acoustic* horizon—that is, the distance that sound waves have propagated since the big bang. This is about a third of the radius of the cosmic horizon itself.) The

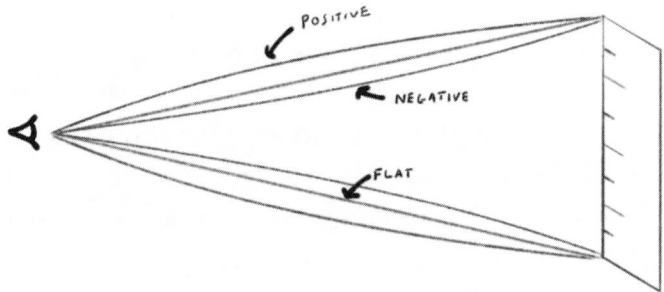

Figure 3.9

A schematic diagram showing the measurement of the curvature from the apparent angular size of a standard ruler. Space-time curvature bends the paths of light rays traveling through space so that the apparent size of an object of known length can be used to measure curvature.

angle subtended on the sky tells us the path taken by light across nearly the entire volume of our observable universe and is a powerful probe of curvature; greater than 1 degree across corresponds to a closed universe, and less than 1 degree corresponds to an open universe (figure 3.9). A careful analysis reveals that the universe is consistent with a flat geometry to within a fraction of a percent.

This remarkably symmetrical result is also a puzzle since from general relativity we know that a flat universe is a dynamically *unstable* state. A flat universe stays flat, but a universe with even a slight positive curvature will become more and more curved as time progresses, and likewise for a universe with a slight negative curvature. Because of this instability, the near-perfect flatness of the universe today means that the universe in very early times must have been flat to an extreme degree—about one part in a million at the time of the last scattering, and at the time of the formation of the first elements when the universe was only a few minutes old, around one part in a trillion. Even the tiniest deviation from flatness would have led to a runaway effect that would have resulted in the universe quickly recollapsing due to positive curvature or expanding into a nearly empty space due to negative curvature. Yet the cosmos we live in is precisely balanced between the two, and the standard big bang model provides us with no explanation why. This is called the *flatness problem* and is a complement to the horizon problem discussed above.

Cosmic Mysteries

In this chapter, we have seen that the LCDM cosmology, despite spectacular success as a general picture of the structure of the universe, leaves many questions of cosmological origins unanswered:

- Homogeneity: The universe is *observed* to be homogeneous to a few parts in a hundred thousand at the time of the formation of the CMB, about three hundred thousand years after the big bang. Yet the size of the cosmic horizon at the time was a little less than a million light-years in radius, subtending an angle of about twice the size of the full moon in the sky today. Points separated by a larger distance were *causally disconnected*; there was insufficient time after the big bang for light to travel from one point to another. Thus the LCDM big bang cosmology provides no explanation for how the universe became so uniform in the first place, and it must be assumed as an initial condition at the beginning of time.

- Flatness: The universe is *observed* to have zero spatial curvature to within a fraction of a percent. Since curvature grows with time, that means the curvature of the universe must have been zero to within one part in a trillion at an age of three minutes, when the first elements were formed. Like homogeneity, this extreme geometric symmetry is unexplained in LCDM and must be assumed as an initial condition.

- Inhomogeneity and cosmic structure: The universe is *observed* to have tiny deviations from homogeneity at the time of the formation of the CMB in the form of growing dark matter halos and acoustic waves in the baryon/photon plasma. These early inhomogeneities formed the seeds of structure in the cosmos today, collapsing under gravity over billions of years to form stars and galaxies arranged in vast cosmic filaments as well as a home to planets, including ours. The LCDM cosmology supplies a detailed and successful picture for the evolution of this structure with time, but it lacks an explanation for the origin of the primordial inhomogeneities.

These unanswered questions do not mean that the LCDM cosmology is incorrect but rather that it is most likely *incomplete*. If we wish to come to a deeper understanding of the structure and origin of the cosmos, we are going to need to find a more comprehensive theory that contains LCDM as a piece. In the next chapter, I introduce the leading candidate for such a theory: cosmological inflation.

4
The Physics of Nothing

Form is emptiness, emptiness is form.
—Avalokiteshvara, *The Prajna Paramita Hrdaya*
(*Heart Sutra*)

Vacuum Energy and Cosmic Acceleration

In chapter 1, I briefly discussed the difference between dark matter and dark energy in terms of their effects on cosmic expansion. The self-gravitation of dark matter acts to slow the expansion of the universe. This is intuitive: if we populate cosmic space with a uniform population of heavy particles, the gravitational force acting between those particles is attractive. Expansion, which pulls the particles apart and dilutes their density, acts against the attractive force of gravity so that the self-attraction of the dark matter "pulls" the universe back together and slows the rate of expansion. Early in the universe's history, the expansion of our universe was slowing down in this way, but starting about eight billion years ago, due to the presence of dark energy, the expansion of our universe started to speed up—an effect we can detect in the Doppler shifts of faraway, ancient supernovas and (indirectly) in the CMB.[1] Dark energy is less intuitive since it behaves as an energy density of empty space and therefore does not dilute as cosmic space expands. If this seems to violate the conservation of energy, that's because it does. Consider a box that expands along with the expansion of the universe, like one of the squares in my expanding grid in figure 2.1. Because the energy *density* of the vacuum is constant, the more the universe expands, the larger the volume of the box gets, and the more energy it contains in total. That energy doesn't come *from* anywhere, any more than the new space inside the box comes from somewhere else; it is being created from nothing as the universe expands. (I will return to this curious question of the creation of new space from nothing in chapter 8 when I discuss quantum gravity and so-called trans-Planckian effects.) The more the universe expands, the more energy is created from nothing, and this acts to push the universe apart so that the expansion speeds up.

Einstein's general theory of relativity enables us to explain this quantitatively. When we apply general relativity to the universe, it tells us that the Hubble constant relating the recession velocity of an object to its distance is proportional to the energy density (of whatever kind) in the space—a law called the *Friedmann equation*. If the energy density is not constant in time, then neither is the Hubble "constant," so a better name for it in general is the Hubble *parameter*. For example, in a matter- or radiation-dominated universe, the Hubble parameter H evolves inversely with time, representing the "slowing down" of the expansion. Because the Hubble parameter is proportional to the inverse of the time, we can express the expansion rate in terms of the time it takes for a region of space to double in linear size:

> The amount of time it takes a ruler expanding along with the universe to double in length is proportional to the inverse of the Hubble parameter, 1/H.

This "doubling time" is also called the *Hubble time*. (Strictly speaking, the Hubble time is the *e-folding* time, in which a comoving length increases in size by a factor of Napier's constant, the irrational number $e = 2.71828$ . . .) The larger the Hubble parameter, the shorter the doubling time. The doubling time of the universe today, based on measurements of the current expansion rate, is about ten billion years.

A universe filled with dark energy behaves very differently from a universe filled with dark matter or radiation. In a universe dominated by the energy of the vacuum, the Hubble parameter is constant and therefore so is the doubling time. The result is *exponential* expansion. Consider again a box expanding in size along with the space in the universe. One doubling time means that the length of each side of the box doubles, which means that the *volume* of the box increases by 2^3, or a factor of 8. After a second doubling time, the volume of the box increases by another factor of 8 so that it is 64 times the original volume. After three doubling times, the volume of the box has increased by a factor of $8^3 =$ 512 relative to its initial size. After 10 doubling times, the volume of a comoving box will have increased by about a factor of a billion! Exponential expansion is the key to solving the cosmic mysteries of flatness and homogeneity discussed in chapter 2.

Inflation

In order to answer the unanswered mysteries of flatness and homogeneity, we need to modify the standard picture of the LCDM cosmology. A key element of the standard LCDM picture is that the energy density of the universe was radiation dominated for about the first sixty thousand years of its evolution, all the way back in time to the moment of the initial singularity. This is, however, an extrapolation. There is good evidence from the primordial abundances of the elements that the universe was radiation dominated at an age of about one second, but the physics of the universe at an age of less than a second is much less certain. As long as we don't alter the behavior of the universe after it is one second old, we are free to modify the simple picture of radiation-dominated expansion without creating conflict with observations. In this spirit, let us consider

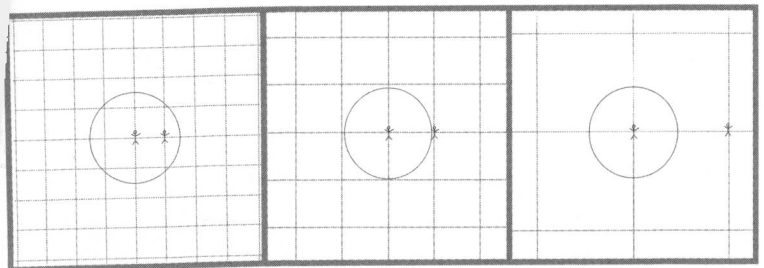

Figure 4.1
The cosmic horizon in inflation. During inflation, the horizon size is nearly constant, but the space is expanding exponentially quickly so that observers initially inside each others' horizon are swept outside and recede from each other faster than light.

is generated in a tiny fraction of a second. This can be visualized another way with a space-time diagram, analogous to figure 3.5. When we add a period of inflation that occurs *before* the onset of the hot, radiation-dominated universe, the initial singularity of the big bang is replaced by the end of inflation, which takes place in the "negative" time beforehand. This is shown in figure 4.2. In this case, the past light cones of distant objects—for example, points on the CMB—overlap and a homogeneous universe presents no causal paradox.

The rapid expansion during inflation also solves the flatness problem in a simple way. Because the cosmic horizon stays at a nearly constant size while the space expands exponentially, any curvature of the initial space-time is stretched to a huge scale, while the observable universe remains tiny. This means that any local patch of the universe appears extremely close to flat, in

the possibility that the universe, when it was much l[...]
than one second old, was dominated by vacuum energ[...]
instead of radiation and as a result underwent a perio[...]
of exponential expansion. Unlike the slow expansion
driven by dark energy today, with a doubling time of
billions of years, this early phase of acceleration would
have occurred when the energy density was extremely
high. In that case, the expansion would have been very
rapid, with a doubling time of a tiny fraction of a second. Such a period of early, rapidly accelerating expansion is called *inflation*.

Accelerating expansion—inflation—is different in important ways from decelerating expansion. Since the Hubble parameter is approximately constant, so is the size of the cosmological horizon. The space inside, though, is expanding exponentially quickly, so that instead of the horizon containing more and more of the universe as time progresses, it contains less and less. Space is swept *outside* the horizon during inflation, as shown in figure 4.1. Two observers initially inside the same cosmic horizon and receding from each other at less than the speed of light are later receding from one another *faster* than light and outside each others' observable universes. This is the key to solving the horizon problem: a tiny patch of the universe, initially in equilibrium, can be stretched to a size far larger than the cosmic horizon. During inflation in the early universe, this process is so rapid that a smooth, equilibrium space the size of our current observable universe

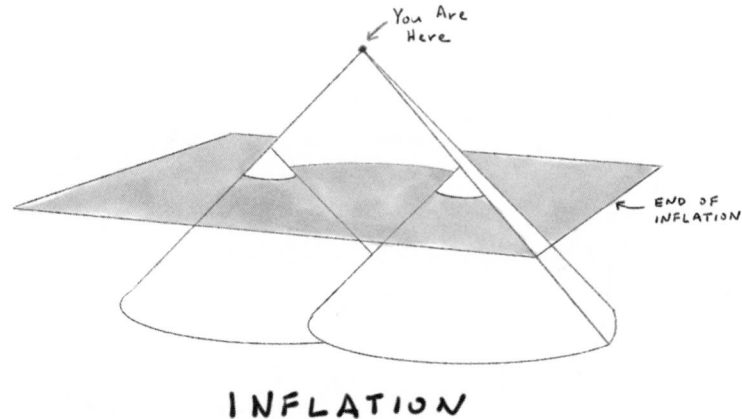

Figure 4.2

A space-time diagram of inflation. The big bang in figure 2.4 is replaced by the end of inflation, and the past light cones of distant objects overlap, eliminating causal paradoxes in a homogeneous universe.

the same way that a small patch of the spherical Earth's surface appears flat. This is shown in figure 4.3. A decelerating universe evolves away from flatness, but an inflating universe evolves *toward* it.

The accelerating expansion of inflation provides a remarkable explanation of a third property of the early universe: the primordial inhomogeneities in the density of the universe at early times. These tiny early fluctuations, which we observe directly in the CMB (figure 3.6), were the seeds for cosmic structure such as galaxies, stars, and planets. Inflation generates primordial perturbations via stretching quantum fluctuations in the vacuum—virtual particles—to sizes far larger than the cosmic horizon. I will discuss this process in detail

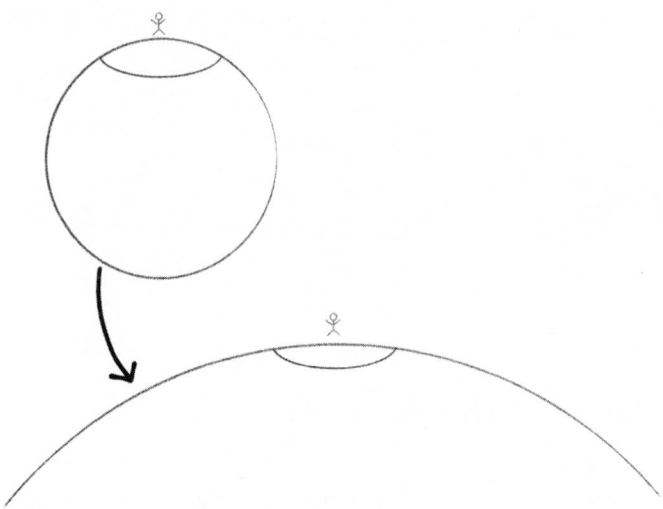

Figure 4.3
The solution to the flatness problem. As a curved surface expands, a small patch appears flatter and flatter to an observer on the surface.

in chapter 5. This inflationary hat trick is compelling: a single postulate—accelerating expansion—ties together and resolves all the cosmic mysteries from chapter 2 at once. Inflation as a solution to the flatness problem is generally credited to a 1980 paper by Alan Guth, who wrote in his personal notes, now in the collection of the Adler Planetarium at the University of Chicago,

> **SPECTACULAR REALIZATION:**
> This kind of supercooling [*inflation*] can explain why the universe today is so incredibly flat—and therefore resolve the fine-tuning paradox pointed out by Bob Dicke in his Einstein Day lectures.[2]

In fact, others were reaching similar conclusions at around the same time, including Russian physicist Alexei

Starobinsky and Japanese physicist Katsuhiko Sato.[3] The idea of inflation was ripe for discovery, and most of the pieces were already in place, but Guth's paper fit all the pieces together and ultimately had the broadest influence. The paper has been cited more than eight thousand times to date.

A key point is that inflation cannot be generated by physics as simple as Einstein's cosmological constant, which postulates a precisely constant energy density for vacuum. The cosmological constant creates the exponential expansion necessary for driving the universe toward a homogeneous, geometrically flat state, but in this case the rapid doubling in size of the universe will continue *forever*. The resulting universe, smooth, flat, and empty, would look nothing like ours. In order to create a universe like our own, inflation must end. This means that the energy of empty space that drives the acceleration must change with time. It must be *dynamical*. It is here that the deep connection between the outer space of cosmology and inner space of particle physics takes root. The physics of fundamental particles provides the perfect mechanism for generating dynamical vacuum energy: symmetry breaking.

Instability and Symmetry Breaking

How do we make the energy of the vacuum in the early universe a *dynamical* thing—that is, one that can change with time? We will utilize a concept that originated in condensed matter physics, was stolen by particle physicists, and that we can now steal again to use in cosmology: symmetry breaking. The basic idea of symmetry breaking is as simple as getting a pencil out of a drawer. Balance the pencil on its tip. What happens? It falls over. This familiar behavior is because the force of gravity always pulls vertically on the pencil so that even a slight tip away from perfectly vertical results in a gravitational force tilting it further still, and the pencil falls. If we apply Newton's laws of motion, we find that a pencil absolutely perfectly balanced on its tip will only fall away from the vertical if it is first tipped away from the state of perfect balance. A pencil perfectly balanced at the vertical is a state of *equilibrium*. The pencil will remain vertical *forever* unless some outside force acts on it. This contradicts our everyday experience, of course, since those outside forces are plentiful: air currents, vibrations, jitters in the atoms in the pencil due to temperature, and even the faint gravitational attraction of surrounding objects. Real pencils, no matter how carefully balanced, fall over almost immediately. This is referred to in physics as an *unstable* equilibrium. The situation where the pencil lies horizontal on the desk is by contrast a *stable* equilibrium; if you pull the pencil up slightly, it will drop back down to horizontal again. Nonetheless, both situations—the pencil precisely vertical and laying at rest on its side—are equilibriums—that is, they are points at which classical physics tells us the pencil will not move at all.

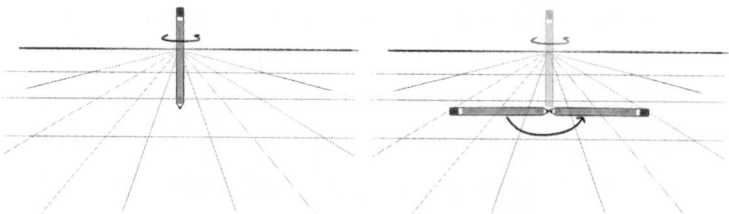

Figure 4.4
A pencil balanced on its tip is symmetric under rotations about the vertical (*left*). The pencil horizontal on the table picks out a direction and "breaks" the symmetry (*right*).

The stable and unstable equilibriums differ in a second key way besides stability: the unstable equilibrium is *symmetric*; we can rotate the pencil about the vertical without changing the system in any way (figure 4.4). Conversely, the pencil on its side—the stable equilibrium—is forced to pick out a particular direction and no longer possesses a symmetry under rotations about the vertical. The symmetry of the unstable equilibrium is *broken*. The balanced pencil is free to fall in any direction, and in perfect equilibrium will do so randomly, but in falling from the unstable equilibrium to the stable one, a particular direction must be picked, and the symmetry is no longer realized by the system. This trivial-seeming phenomenon, referred to as *symmetry breaking*, is a deep feature of fundamental physics, applying in systems as divergent as ferromagnets and the Higgs boson. For example, in a ferromagnet, the symmetric state is the unmagnetized state in which the magnetic fields of the constituent atoms are randomly arranged, and the system picks out no preferred direction in space. The symmetry breaking state is the one for which the magnetic fields of the atoms are aligned, picking out a particular direction in space and making a magnet. Like the pencil, ferromagnets can magnetize equally easily in any direction because of the underlying rotational symmetry of the system.

We can diagram the falling pencil by plotting the *potential energy* of the pencil in various angles of the tip. Any object lifted against gravity will have potential energy, which is released and converted into kinetic energy as it falls in the gravitational field. In the case of the pencil, the potential energy is highest when it is in the unstable equilibrium, balanced on its tip, and lowest in the stable equilibrium, laying on its side. The potential energy depends on the angle of the tip of the pencil, but not on the *direction*, because of rotational symmetry. If we plot the potential energy as a function of the orientation of the pencil, it takes a shape reminiscent of a sombrero, with a hilltop in the center surrounded by a circular minimum representing a symmetric state with the pencil laying on its side (figure 4.5)—hence the common name "Mexican hat." The falling of the pencil is like a ball rolling off the top of the central hill and coming to rest at the minimum of the potential, represented by the circular trough at the bottom.

The usefulness of the potential surface as a representation of symmetry breaking is that it can represent *any*

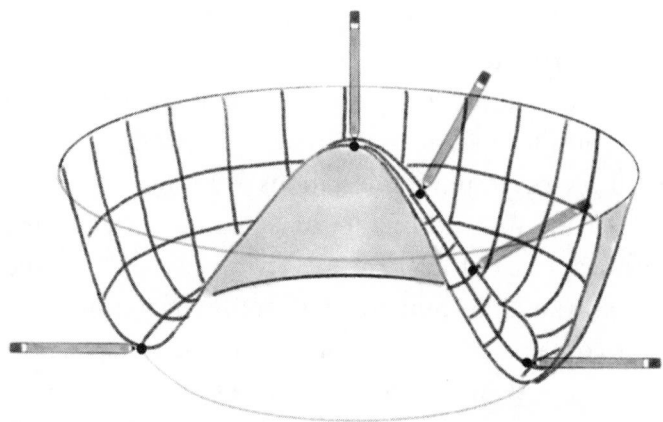

Figure 4.5
A cutaway view of the potential energy of a falling pencil as a function of the angle of the tip. Symmetry breaking is analogous to a ball rolling off a hill. The symmetric equilibrium of a pencil balanced on its tip is at the peak, and the stable equilibrium with the pencil on its side is the circular trough of the sombrero.

system evolving from an unstable equilibrium to a stable one, not just the example of the balanced pencil. For instance, in a ferromagnet, the peak of the potential is the unmagnetized state, and the minimum is the state with the atoms aligned and material magnetized. The position in the circular trough, like the orientation of the pencil, represents the direction of magnetization. The representation of a ball rolling off a hill applies in more abstract cases as well, such as the Higgs boson in the standard model of particle physics, first detected at CERN in 2012.[4] Symmetry breaking is key to the function of the Higgs in the standard model, which is to give mass to the other particles in the theory, such as quarks and the force-mediating particles, the W and Z bosons. In the case of the Higgs, the symmetry is not with respect to physical rotations as in the instance of the balanced pencil or a ferromagnet but rather in the abstract space of field theory. The principle, however, is identical: the symmetric state of the Higgs is unstable and contains potential energy, while the stable state breaks the symmetry and minimizes the potential. In this way, the Higgs boson, like the balanced pencil or ferromagnet, can be represented by a ball rolling off the peak of a potential like that in figure 4.5. When the ball is balanced at the peak, the Higgs state is symmetric, and all the particles of the standard model are massless. When the ball rolls to the minimum of the potential surface, it picks out a direction and breaks the symmetry, and the particles of the standard model acquire mass in proportion to the strength of their interaction with the Higgs. Nevertheless, there is a crucial difference between the Higgs and a pencil or ferromagnet: the Higgs is a *field*, which means it takes a value at every point in space. The Higgs field is remarkably reminiscent of the nineteenth-century concept of the luminiferous aether: it permeates all of space, and interaction with standard model particles as they move in the medium of the Higgs field is the source of inertia and hence mass. Think of the symmetric state of the Higgs field as an *infinite* number of pencils balanced on their tips, one for each point in three-dimensional space, and when the unstable Higgs symmetry breaks, all the pencils fall at once (figure 4.6). This magnificent idea of symmetry breaking as the source of mass, known as the *Higgs mechanism*, earned François Englert and Peter Higgs the Nobel Prize in 2013.[5] (A third originator of the Higgs mechanism, Robert Brout, died in 2011, a year too soon to see his idea vindicated by discovery.)[6] The particle associated with the Higgs field, the Higgs *boson*, is the quantum state associated with oscillations of the field about the stable minimum.

The property of the Higgs field that it permeates all of space—all of the *universe*—is where it makes a connection with cosmology. Remember that the symmetric state of the Higgs, like the pencil balanced on its tip, carries potential energy. Since the Higgs permeates all of space, so too does its potential energy; the Higgs field in its unstable state acts exactly like a cosmological constant! This remarkable connection between the Higgs boson and an effective cosmological constant was first proposed in 1974 by Russian physicist Andrei Linde and independently by US physicist Joseph Dreitlein.[7] Furthermore, since the Higgs is dynamical, so too is the associated vacuum energy. The Higgs field, or another field with similar properties, is exactly what we need to provide the time-dependent vacuum energy necessary for inflation; the vacuum energy originates in the unstable symmetry, and the dynamics is provided by the breaking of that symmetry.

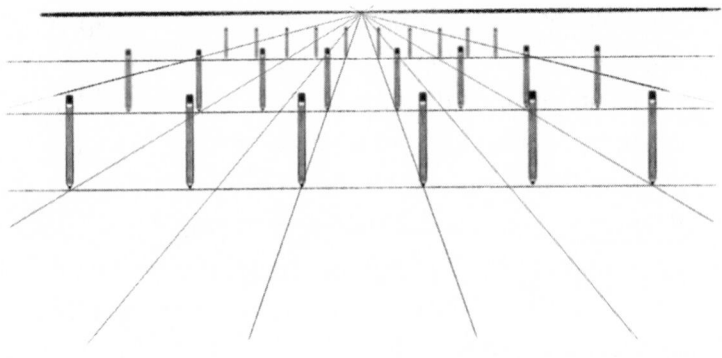

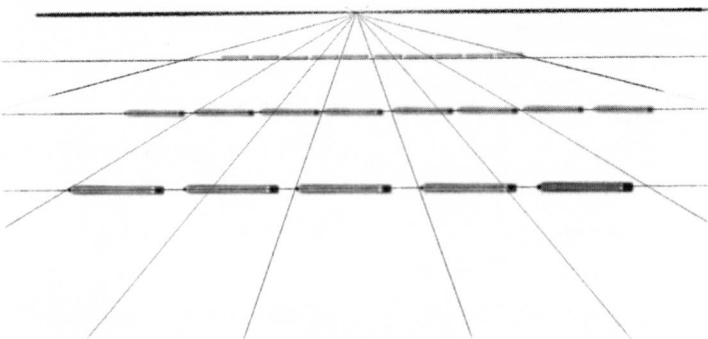

Figure 4.6

A metaphor for the Higgs field. Symmetry breaking in the Higgs field takes place at every point in space. The top panel shows the unstable symmetric state, and the bottom panel depicts the stable state with broken symmetry.

The Inflaton

The physics of symmetry breaking provides us with exactly the properties required for inflation to happen in the early universe. Symmetry breaking is also ubiquitous in nature, realized in systems as diverse as mechanical ones (like the example of a pencil), phase transitions in solids and liquids, and fundamental particles like the Higgs boson. The basic model for symmetry breaking in inflation is, similarly, a fundamental field like the Higgs, which undergoes symmetry breaking—an unstable symmetric equilibrium decaying into a stable state with broken symmetry—which can be modeled by exactly the same sombrero-like potential surface as a pencil falling from the vertical, or the Higgs field. The identity of this field responsible for inflation is unknown and referred to generically as the *inflaton* field. The inflaton could even be the standard model Higgs, but it need not be, since *any* symmetry breaking transition will have the same physics.[8] Inflation could be driven by a fundamental field like the Higgs, a composite of many other fundamental fields, a phase transition akin to the onset of boiling in a fluid, or gravitational effects arising from modifications to the standard theory of general relativity. (For instance, in chapter 7, I will explore the possibility of the inflaton field arising from the way in which extra dimensions in string theory are "rolled up" into compact spaces.) Moreover, the idea of a potential surface is more general than the particular shape of a

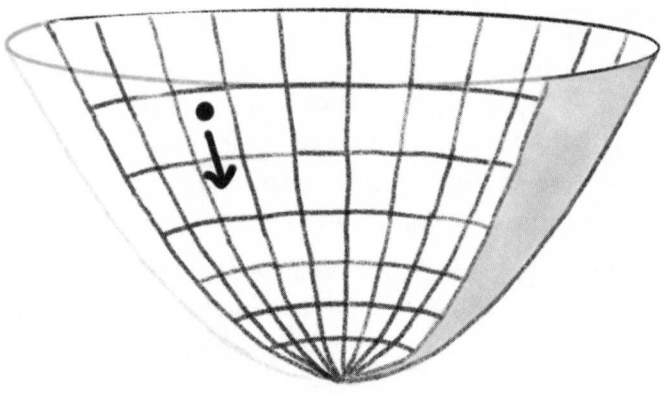

Figure 4.7
A potential surface with no symmetry breaking.

sombrero; the potential surface for a field can in principle take *any* shape as long as there is at least one minimum to define a stable state. The simplest of these is a
bowl shape (figure 4.7), in which the symmetric state
is the minimum of the potential surface, and spontaneous symmetry breaking does not occur. An analogy
for this would be a pendulum for which the symmetric position—hanging vertically downward—is stable.
In what follows, it is important to keep in mind that
the "inflaton" field is a stand-in for entirely unknown
physics.

Understanding symmetry breaking in cosmology
necessarily includes general relativity since the structure
of the cosmos is set by gravity. We have already encountered one such novel effect: vacuum energy induces
accelerated expansion, the source of Guth's "spectacular

realization" in 1979. But there is another question that must be solved: How do we make inflation last long enough to serve as a solution to the unanswered questions of the LCDM cosmology? The universe during inflation must double in size about eighty times at least—a factor of 10^{24}—to explain the global properties of the universe. Guth solved this problem by suggesting that the symmetric state of the inflaton is *metastable*. Think of a metastable point as a dimple in the peak of the potential so that the inflaton is "stuck" for long enough for inflation to smooth and flatten the universe (figure 4.8) before finally decaying to the stable minimum. Guth realized at the time that this idea "leads to some unacceptable consequences"—in particular, "the central problem is the difficulty in finding a smooth ending to the period of exponential expansion."[9] For a time, this was considered an unsolvable problem.

The problem was not unsolvable, however, and two solutions were proposed. The first, in the form of evolution from an unstable equilibrium called "new inflation," was proposed in a paper by Linde, and later in a paper by Andreas Albrecht and Paul Steinhardt.[10] The two papers were published in early 1982, less than three months apart. The second solution came in a model by Linde in 1983 for which the inflaton evolves in a potential like that portrayed in figure 4.7, toward a minimum state with no breaking of symmetry at all, called "chaotic inflation."[11] The scenarios differ in their initial conditions, but both models rely on a general relativistic

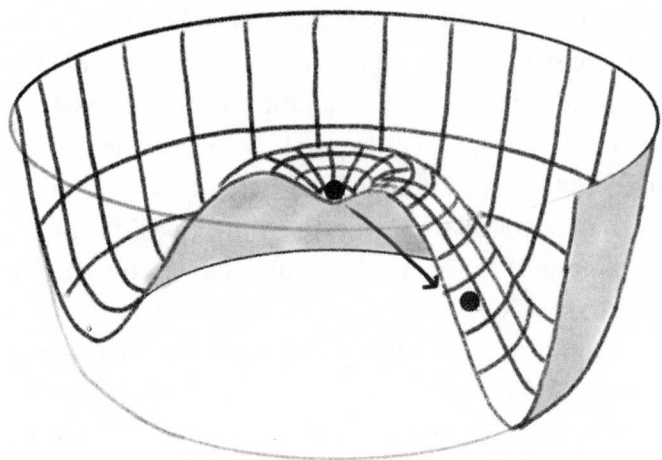

Figure 4.8

Guth's model of inflation: a potential surface with a metastable minimum at the symmetric point on the hilltop and a stable minimum that breaks the symmetry. Classically, the system will remain at the metastable point forever, but quantum mechanics allows it to "tunnel" directly from the metastable symmetric point to the stable minimum.

effect to ensure that inflation lasts long enough to explain the observed flatness and homogeneity of the current universe: drag from cosmological expansion. We encountered expansion drag in chapter 2: any object moving relative to the cosmic rest frame will eventually slow and come to rest relative to the expansion. The key point is that expansion drag applies to *any* dynamical process, including the evolution of the inflaton field. In the early universe, when the expansion rate was dozens of orders of magnitude larger than it is today, the effect was substantial. Think of symmetry breaking in the presence of expansion drag as a pencil falling downward from balance inside a jar of thick molasses; friction from the viscous fluid will slow the fall of the pencil. Consider for simplicity a universe whose only constituent is the energy from the inflaton field, which we represent as a ball rolling on the hill of the potential surface. The energy of the field has two pieces: the first is the potential energy, and the second is the *kinetic* energy associated with the ball's motion on the hill. The key difference is that expansion drag behaves as a frictional force and acts on the kinetic energy, but not on the potential energy. The field rolls more slowly. If the frictional force is large enough, and the field rolls slowly enough, the field behaves effectively as a cosmological constant and drives the accelerated expansion of the universe. This type of evolution is referred to, naturally enough, as *slow roll*.

Two factors control the slow rolling of the inflaton: the height and steepness of the potential. The steeper the potential surface is, the faster the field rolls, so slow roll inflation requires a gently sloping, or "flat" part of the potential surface. In a standard symmetry breaking potential like that of figure 4.5, this means that inflation takes place when the field is near the symmetric point at the top of the hill. The height determines the potential energy, which in turn sets the expansion rate of the universe. The higher the potential, the higher the expansion rate, and the stronger the drag on the inflaton, which slows the rolling. A higher expansion rate also means that the universe inflates more quickly, doubling in size in a shorter time. This means that inflation does its job better as the energy involved becomes higher; the exact value is model dependent, but typically inflation requires an energy of around 10^{15} *billion* electron volts, or about a hundred billion times the energy of the LHC and far beyond the energy of any known physics. This is difficult to convert into everyday units in a meaningful way; the density of the universe during inflation would have been about 10^{77} grams per cubic centimeter, or more than 10^{60} times the density of an atomic nucleus. At the time, the radius of the cosmic horizon would have been about the size of a grapefruit, or Lemaître's primeval atom.[12]

Slow Roll and Reheating

I am now in a position to sketch out a basic model for inflation that captures all the elements necessary to provide an explanation for the cosmic mysteries outlined in chapter 2. I will consider the case of a single inflaton field describing some sort of symmetry breaking process in the early universe. I visualize this as a ball rolling off the top of a hill, slowed by friction from the rapid expansion. The steeper the hill, the faster the ball rolls, and the shorter the duration of the inflationary epoch. We need inflation to last long enough—at *least* eighty doubling times—which is easy to accomplish if

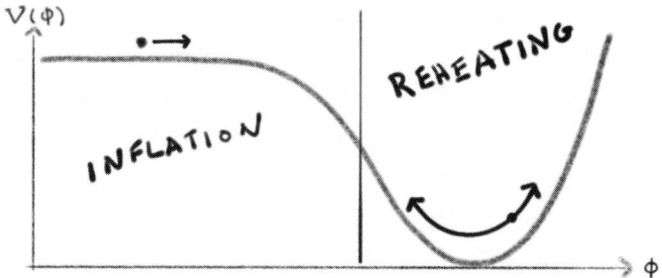

Figure 4.9

A schematic representation of the potential surface for a simple slow roll inflation scenario. Accelerated expansion (inflation) occurs while the field slowly rolls on the flat portion of the potential. Once the potential steepens, inflation ends, and the field oscillates about the minimum. Interactions of the inflaton with other particles cause the universe to "reheat" to a hot thermal state.

the potential V plotted as a function of the field ϕ is flat near the top, as shown schematically in figure 4.9.

The universe during slow roll inflation is very different from the hot, thermal equilibrium state of the standard big bang picture. For one thing, it is not hot! Any matter or radiation present in the universe at the onset of inflation will be diluted away exponentially quickly, so that after a few doubling times, most cosmic horizons will contain no particles at all, and the temperature is driven rapidly to zero. Likewise, if the matter in the universe is not perfectly smooth to begin with, it very rapidly becomes so. Shortly after inflation sets in, the universe approaches an extremely simple state of almost perfect homogeneity and temperature of nearly absolute zero, described by a coherent quantum state called a *Bose-Einstein condensate*. Nothing is left in the universe but the field and its potential energy, driving accelerating expansion. It is in fact this emptying of the universe by inflation that was one of Guth's original motivations for proposing the theory. Models popular at the time for unifying the strong, weak, and electromagnetic forces in a single grand unified theory came with the side effect of predicting the copious production of magnetic monopoles in the very early universe—as pointed out by theorist Tom Kibble in 1976.[13] Such monopoles—a particle corresponding to a magnetic north or south pole with no opposite pole attached—are not observed to exist in nature. Inflation gets rid of these unwelcome visitors by diluting their density exponentially close to zero.[14] In theories such as supersymmetry, inflation removes other unwanted relics such as particles called gravitinos.[15]

Inflation doesn't just remove unwanted relics. It removes *everything*. A zero-temperature Bose-Einstein condensate hardly resembles anything like the hot, thermal equilibrium state that we know must have existed at least by the time the universe was a few seconds old and the nuclei of the primordial elements began to form. Any successful theory of inflation must explain the transition between the two. This was the Achilles' heel of Guth's original theory (now called "old inflation"), which was unable to transition to a homogeneous thermal equilibrium state because the inflaton is trapped too effectively in the metastable point at the top of the potential (figure 4.8) and forms a froth of disconnected bubbles instead of a smooth equilibrium state. Guth in fact pointed out this flaw in his original paper: "The above statements do not quite prove that the scenario is impossible, but these consequences are at best very unattractive. Thus, it seems that the scenario will become viable only if some modification can be found which avoids these inhomogeneities. Some vague possibilities will be mentioned in the next section."[16] This shortcoming was rectified in 1982 by Linde, and independently by Albrecht and Steinhardt.[17] Both papers proposed a type of single-bubble or "new" inflation, in which the field rolls away from an *unstable* symmetric point and undergoes slow roll. The transition from the inflationary state to the hot thermal equilibrium one of the big bang was described in 1982 by Albrect and Steinhardt along with Michael Turner and Frank Wilczek, and is referred to as *reheating*.[18]

In Guth's original model, inflation ends when the field quantum mechanically "jumps" or *tunnels* out of the metastable symmetric point of the potential. In the later "new" inflation model, the field rolls slowly off the hilltop, slowed by expansion drag, with no sudden jumps. As long as the potential surface is flat enough, slow roll continues, and the universe continues to inflate, emptying the space as well as driving it toward flatness and large-scale uniformity. Yet this does not continue forever; eventually the potential steepens, and expansion drag is no longer sufficient to keep the field slowly rolling. At this point the inflaton rapidly drops to the minimum of the potential surface and inflation ends. The vacuum energy present during inflation does not go away, however. Instead, the potential energy of the field in the unstable state is turned into *kinetic* energy near the minimum, and the field oscillates about the minimum. This process is depicted schematically in figure 4.9. During this oscillating phase, the expansion no longer accelerates, but the universe is still cold, nearly at absolute zero temperature. If the inflaton field is alone and interacts with no other particles, this is the end. The universe gradually expands, the oscillations slow due to expansion drag, and the field settles into its minimum, leaving behind a cold, empty universe entirely unlike our own. But if the inflaton field *interacts* with other particles such as those of the standard model, something else altogether happens. The oscillating field decays into the particles to which it couples, which then interact with one another, creating still more particles and turning the kinetic energy of the oscillating inflaton field into radiation, heating the universe from zero temperature to billions of degrees. The details of this "reheating" process depend mostly on unknown physics—neither the identity of the inflaton field nor the particles to which it couples are understood—and the process is rich with complex possible physics. (For modern reviews, see "Further Reading.")[19] No matter what path it takes, however, the end point is the same:

a hot, thermal equilibrium universe, exactly as required to set the initial conditions for the big bang.

The description of inflation as being driven by a "field," the inflaton, makes it possible to add dynamics to the basic picture of accelerating expansion driven by vacuum energy. In the slow roll picture, the vacuum energy is temporary because the field is evolving from a symmetric maximum to a minimum with broken symmetry, like the analogy of a pencil falling away from balance on its tip. Like any good scientific model, this field picture does more than allow us to answer the original question of what causes the universe to approach a flat, homogeneous state. It lets us ask—and answer— new questions as well, such as, How does the universe end up in a hot thermal state? The field picture of inflation, it turns out, is more powerful still. It does not just explain why the universe is flat and uniform but also provides a remarkable explanation of the initial, tiny *non*uniformity of the universe too. This process, closely related to Hawking radiation from black holes, is the subject of the next chapter.

5
The Quantum Vacuum and Cosmic Structure

"Take some more tea," the March Hare said to Alice, very earnestly.

"I've had nothing yet," Alice replied in an offended tone, "so I can't take more."

"You mean you can't take less," said the Hatter: "it's very easy to take more than nothing."

"Nobody asked your opinion," said Alice.

—Lewis Carroll, *Alice's Adventures in Wonderland*

It may seem puzzling at first to hear cosmologists speaking of the universe using terms like "smooth" or "homogeneous" when the world we see around us is anything but. Our local universe is highly *inhomogeneous*, with dense planets separated by vast gulfs of nearly empty space. Our galaxy roils with clouds of dust and spiral arms, centered on a galactic core that houses a black hole of more than a million solar masses, condensed into a space only about a quarter the size of the orbit of Mercury. Widening our view further, we find that the universe on larger scales consists of galaxies much like our own—island universes separated from one another by millions of light-years of empty space. Even these galaxies are clustered into larger structures still; the Milky Way is a member of a gravitationally interacting group of about a hundred thousand galaxies known as the Laniakea Supercluster, extending about 500 million light-years across. Clearly, the idea of a homogeneous cosmos implied by the cosmological principle is in some sense an idealization.

The idealization is a statistical one. Our current universe *is* homogeneous—at least in an average sense—if you look at it with a wide enough lens. Surveys of the distribution of galaxies in the universe show that the scale at which the universe becomes homogeneous on average is around 325 million light-years, years, making our local supercluster a bit of an outlier, but not incompatible with this number.[1] This means that if we blur our vision so that the smallest things we can see are between 300 and 500 million light-years across, every part of the universe looks pretty much like every other part. Furthermore, we know that the inhomogeneity of the universe is growing with time because of gravity. If we look out in space—and back in time—we see that the universe far away, and therefore earlier in its history, was smoother than it is today. When we look all the way to the edge of the observable universe in the CMB, we observe that the universe less than a million years after the big bang was very smooth indeed. Fluctuations

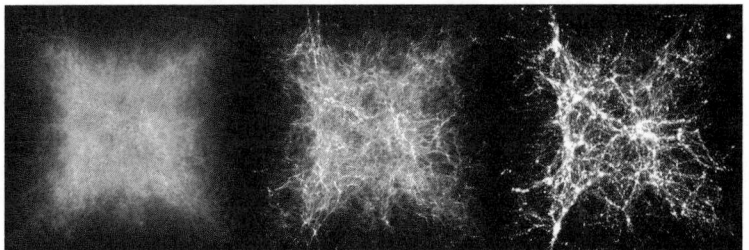

Figure 5.1
Primordial density perturbations in the early universe collapsing under gravity to form the cosmic web of structure seen in the universe today. Images: GADGET2 computer simulation by Leigh Korbel and the author.

in the temperature of the CMB, which map fluctuations in the density of the universe at the time of recombination, were a few parts in a hundred thousand, meaning that the universe at that time was homogeneous to within a few thousandths of a percent.

We have a good idea of how this early, almost perfectly smooth state came to look like our lumpy, inhomogeneous universe today: gravity. The physics is simple. Regions that are denser than their surroundings tend to collapse under their gravitational self-attraction, and regions that are less dense get emptied out and form voids because of matter falling onto the dense regions. This is a simple rich-get-richer scenario by which the tiny ripples in the density of the early universe eventually collapse into the dramatic "cosmic web" of filaments and voids we see today in the distribution of galaxies (figure 5.1). Dark matter is the key player in this process. Without it, there would be little structure in the universe at all because the initial dense regions would be erased by pressure from photons before they ever had a chance to collapse.[2] For this reason, cosmic structure is a powerful tool for understanding the properties of the dark universe, which can (so far) only be detected via its indirect effect on the luminous parts of the cosmos. As important as this process is to the present universe, our question here is a different one: What set the *initial conditions* for the vast cosmic dance of structure formation? Where did the initial tiny fluctuations in the density of the universe come from? Inflation provides a remarkable, elegant answer: quantum uncertainty. We saw in the last chapter that the end point of inflationary expansion is a completely empty, zero-temperature universe, which is by definition completely smooth. Whatever was in the universe before inflation was diluted away by expansion, leaving nothing behind. But "nothing," it turns out, is a complicated thing.

The Quantum Vacuum

In the universe of Newton and Descartes, empty space was just that: formless and absolute. Newton wrote at the beginning of his opus *Philosophiae Naturalis Principia Mathematica*, "Ut partium Temporis ordo est immutabilis, sic etiam ordo partium Spatii," or "As the measure of time is immutable, so is the order of space."[3] The idea of space as an unchanging arena for dynamics is fundamental to Newtonian physics and held force until the twentieth century, when it was upended by two monumental ideas. The first was Einstein's general theory of relativity, which transformed space itself into a dynamical entity, capable of both being warped by the material bodies within it and changing with time. In the words of physicist John Archibald Wheeler, "Space tells matter how to move. Matter tells space how to curve."[4] The second was the Heisenberg uncertainty principle, which postulated that localizing a particle in space is accompanied by an unavoidable uncertainty in its momentum.[5] It is not obvious what this has to do with the structure of space, but quantum uncertainty fundamentally alters our definition of a vacuum—a space containing no particles. Nothing, it turns out, is very much something.

Consider a hydrogen atom—an electron in orbit around a proton. A basic problem of atomic structure at the dawn of the twentieth century was how such a system could be stable. Classical physics provided no explanation. Maxwell's laws of electromagnetism implied that the moving electron would radiate, losing energy in the process, and spiral into the proton. Quantum uncertainty prevents this; the closer we squeeze the electron to the proton at the center of the atom, the higher the uncertainty in its momentum, which carries the electron away again. This can be seen quantitatively by writing the Heisenberg uncertainty principle relating the radius r of the electron orbital and its momentum p to Planck's constant $\hbar$:

$$\Delta p \Delta r = \hbar.$$

We can estimate the radius of a hydrogen atom by calculating the uncertainty

$$\Delta r = \hbar \,/\, \Delta p \,,$$

where the uncertainty in the momentum can be estimated from Coulomb's law for the energy of the electromagnetic interaction between the electron and the proton,

$$E = \frac{p^2}{m_e} = k_e \frac{e^2}{r}.$$

(Here e is the electromagnetic charge of the electron, m_e is its mass, and k_e is the universal constant determining the strength of the electromagnetic force, and we estimate $\Delta r \sim r$ and $\Delta p \sim p$.) Combining these results in the relation

$$\Delta r \sim \frac{\hbar^2}{k_e e^2 m_e} \sim 10^{-10}\,\text{m}.$$

Thus the smallest we can squeeze the electron orbit around the proton is around 10^{-10} meters, which is in fact a good estimate of the size of a hydrogen atom. This is a simple example of *zero-point energy*: the lowest possible energy for the electron in a hydrogen atom differs from zero because of quantum uncertainty.

The zero-point energy created by quantum uncertainty applies to *all* systems, no exceptions. This includes fields like the Higgs and inflaton, which permeate all space. Imagine a microscope of hypothetically infinite magnifying power; as you use the microscope to probe shorter and shorter distance scales, the associated uncertainty in momentum (and therefore energy) becomes larger and larger, eventually exceeding the mass (via $E = mc^2$) of the Higgs boson, or any other particle in the standard model. Once the uncertainty in energy exceeds the threshold of twice the particle mass, it becomes possible to create—out of nothing—a particle/antiparticle pair. These quantum fluctuations of the vacuum are known as *virtual* particles. Even a perfect vacuum is not empty but instead is populated, via quantum uncertainty, with an infinite depth of *virtual* objects, whose existence only becomes definite when probed at sufficiently short distances or (equivalently) high energy. This is not philosophy; the effects of vacuum fluctuations are measurable in real physical systems. For example, correctly calculating the properties of particle collisions at accelerators like the LHC requires taking into account virtual interactions. Another illustration is the *Casimir effect*, which is an attractive electromagnetic force between two flat metal plates in a vacuum due to virtual particle interactions.[6] The existence of this force has been verified in multiple experiments.[7] Yet another observable effect is *vacuum birefringence*, in which virtual particles induce a polarization into light traveling through a vacuum. Indirect evidence for vacuum bire-fringence was observed in a neutron star system in 2017, and efforts are underway to directly detect the effect in laser systems.[8]

When we consider the quantum nature of the vacuum in Einstein's general theory of relativity, we discover new effects. A possibly familiar example is Hawking radiation of a black hole; an observer at an infinite distance from a black hole in vacuum will see the black hole radiating particles in thermal equilibrium, with a temperature inversely proportional to the mass of the black hole.[9] A cartoon explanation of Hawking radiation is virtual pair creation near the event horizon of a black hole. One half of the particle/antiparticle pair falls into the black hole horizon, never to be seen again, and its orphaned partner escapes to infinity as Hawking radiation. In 1977, Gary Gibbons and Hawking proposed that this process applied not only to black hole horizons but *cosmological* horizons as well.[10] Within two years of Guth's proposal of inflation in 1980, Hawking, and independently Guth, Starobinsky, and James Bardeen, Steinhardt, and Turner, applied the result to inflationary cosmology.[11] In the rapidly doubling spacetime of inflation, virtual pairs appearing from the vacuum are swept apart, outside their respective cosmic horizons, before they have a chance to recombine and disappear. The result is that the cosmic horizon in inflation, like the horizon of a black hole, radiates.

Keep in mind that the cartoon picture involving virtual particle pairs is just that: a cartoon. A more accurate way to think of it is that in general relativity, the quantum mechanical definition of empty space is *observer dependent*. The simplest example of this (although not the first discovered) is the *Unruh effect*, proposed by William Unruh in 1976.[12] Unruh showed that an observer in an accelerated reference frame in a vacuum would observe virtual particles as *real* ones in a thermal distribution with finite temperature. All inertial (i.e., unaccelerated) observers agree on the definition of a vacuum, but noninertial observers do not. Similarly in the case of a black hole, it is impossible to define what you mean by a vacuum everywhere in space at once. A "vacuum" as defined by an observer near the event horizon of a black hole appears to be a source of radiation to an observer looking down on the black hole from a distance. Likewise, what appears to be empty space to one observer in an inflating space-time is no longer empty from the point of view of another observer, outside that observer's cosmic horizon. This malleable nature of who sees a vacuum and who does not is the key to how inflation sets the initial condition for the formation of structure in the universe.

Primordial Perturbations

The key property of inflationary space-times is the exponentially rapid stretching of space, which carries quantum fluctuations along for the ride. Just as we can describe light either as a particle (photons) or wave,

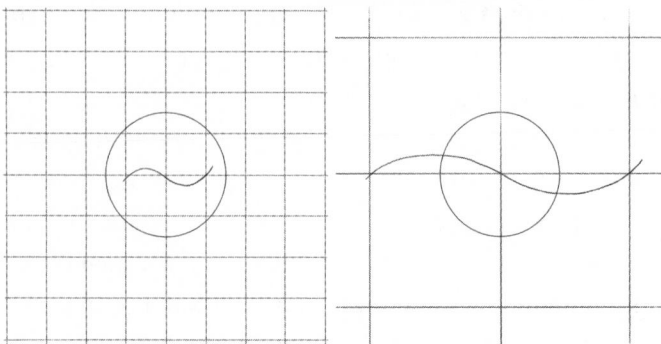

Figure 5.2
Quantum modes with a wavelength smaller than the cosmic horizon size are stretched by inflation to wavelengths larger than the horizon.

virtual "particles" can be described using their wave properties. Just as photons increase in wavelength with expansion, so do virtual quantum fluctuations. In inflation, this redshift happens exponentially quickly so that any wave with a wavelength much smaller than the size of the cosmic horizon is almost instantly stretched to a wavelength *larger* than the cosmic horizon (figure 5.2). In the particle picture, this corresponds to the virtual particles being swept outside each others' horizons so they cannot recombine. In the wave picture, once the virtual mode is stretched to larger than the horizon size, it is no longer virtual; it is a real, classical wave. Since quantum modes are constantly being stretched by the inflationary expansion, these *superhorizon* waves are continuously generated, from empty vacuum, by quantum effects closely related to those that produce Hawking radiation in black holes.

Waves with wavelengths larger than the horizon do not behave like ordinary waves. A normal wave—for instance, propagating light waves or waves on water—oscillates with a characteristic frequency that is related to its wavelength. The shorter the wavelength, the higher the frequency. Superhorizon waves, by contrast, don't oscillate at all but instead "freeze" as their wavelength stretches past the horizon. Once the wave is frozen, its amplitude—the height of the wave—remains constant. During inflation, the size of the cosmic horizon is nearly constant. After the universe reheats, the cosmic horizon grows and catches up with the quantum modes that were stretched outside the horizon during inflation. Those modes then reenter the horizon as classical perturbations. The wave *amplitude* (its height) can be calculated using quantum field theory, with the result that any such frozen mode has an amplitude proportional to the expansion rate H. The higher the expansion rate (and correspondingly, the higher the vacuum energy during inflation), the stronger the superhorizon waves generated. This process of generating modes from a vacuum, which stretch to a superhorizon size and freeze with an amplitude proportional to H, applies to *all* scalar fields—for example, the Higgs boson as well as the inflaton field itself. For most fields, these stretched vacuum fluctuations are too small to be detectable, with the exception of fields that couple to gravity—that is, the shape of spacetime itself. In the simplest inflation models, there are two such fields: gravitational waves and the inflaton.

Let us consider the simpler case of gravitational waves first. A gravitational wave is, as the name suggests, a wave propagating via gravitational fields in much the same way that an electromagnetic wave (light) is a wave propagating via electromagnetic fields. For example, a radio transmitter creates electromagnetic waves via electric currents in an antenna. A transmitter takes charged particles (electrons) and shakes them back and forth, and the accelerating charges create waves in the associated electromagnetic fields. Similarly, if one shakes a massive object back and forth, it creates waves in the gravitational field generated by the object. For objects of ordinary mass, these waves are far too small to detect, but in the case of mergers of black holes of a few dozen solar masses, these gravitational ripples can be seen by instruments such as LIGO/Virgo, and were first detected on September 14, 2015, receiving the Nobel Prize in 2017.[13] Since gravity as described by general relativity is just the curvature of space-time, gravitational waves can be thought of as waves in the shape of space-time itself, and it is expected that quantum mechanics applies to these waves just as for any other. Just as electromagnetic waves have a particle counterpart—the photon—gravitational waves have an associated particle—the *graviton*. Gravitons, like photons or the Higgs boson, have zero-point fluctuations, and exist in a vacuum as virtual particles that can be stretched and frozen by inflationary expansion. This is a *generic* prediction of inflation: gravitational waves are generated by any accelerating expansion, regardless of the identity of the field (or fields) responsible for the acceleration. The higher the energy scale of inflation, the faster the expansion and stronger the gravitational waves produced. If the energy scale of inflation is high enough—around 10^{15} billion electron volts—these gravitational waves will be detectable via their effect on the CMB. I will discuss current observational bounds in chapter 6.

The second field that couples to gravity is the inflaton itself because it is the energy of this field that drives inflation. Quantum fluctuations in the inflaton are the same as for any other field, but the way they interact with gravity is unique. Let us revisit our picture of slow roll inflation as a ball rolling down the hill of the potential surface created by symmetry breaking. The energy density of the universe is set by the ball's position on the hill: the higher up the hill, the higher the energy, and the faster the expansion. Slow roll is evolution according to classical physics, so that the ball rolls always down the slope. When we include fluctuations due to quantum uncertainty, however, this adds a random component to the ball's roll. Quantum effects cause the ball to jump randomly, both up and down the slope (figure 5.3). Since the position on the potential surface determines the energy density, an upward jump corresponds to an increase in density and therefore the expansion rate, while a downward jump corresponds to a decrease. In this way, quantum fluctuations in the

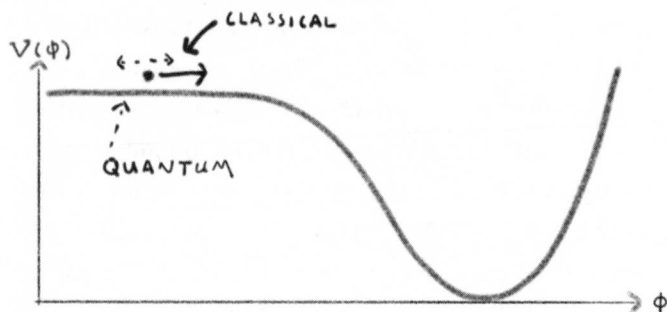

Figure 5.3
As the inflaton field rolls down the potential surface (classical evolution), quantum fluctuations add random "noise" to the motion,
either up or down the potential.

inflaton field generate fluctuations in the *density* of the
universe.

A detailed calculation of the density perturbations
generated during inflation requires quantum field theory, but the result is simple to state: the deviation $\delta\rho$
from the average density, in units of the average density $\bar{\rho}$, is given by the ratio of the size of the quantum
fluctuations to the distance that the field rolls along the
potential in one doubling time of the expansion,

$$\frac{\delta\rho}{\bar{\rho}} = \frac{\langle \delta\phi \rangle_{\text{quantum}}}{\langle \delta\phi \rangle_{\text{classical}}}.$$

The amplitude of the quantum fluctuations is determined by the expansion rate: the faster the expansion
(and shorter the doubling time of the universe), the
larger the quantum fluctuations. The rate at which the
field rolls along the potential surface is determined by

two factors. The first is the expansion rate: the higher the expansion rate, the *slower* the rolling of the field. The second factor is the steepness of the potential: the steeper the hill of the potential surface, the faster the field rolls, which makes the resulting density perturbations smaller. Since the size of the density perturbations produced depends on the *ratio* of the rates of quantum and classical evolution, it likewise depends on both the expansion rate and slope of the potential surface. The quantum fluctuations are typically much smaller than the corresponding classical rolling, so the corresponding density perturbations are also small. (In chapter 7, I will look at what happens when quantum fluctuations dominate over classical evolution.)

Since the quantum jumps in the field are completely random, so too are the resulting density perturbations. Quantum mechanics makes a definite (but statistical) prediction about this randomness: the random fluctuations created form a *normal* or *Gaussian* distribution, also known as a "bell curve." Examples of Gaussian distributions are abundant in nature, such as the distribution of shoe sizes in humans or short-term variations in stock prices. A simple example of a probability distribution that approximates a Gaussian is the roll of pairs of dice. There are thirty-six possible outcomes. Of those thirty-six possible outcomes, there are six ways of rolling dice that total seven, but only three ways of rolling dice that total four (figure 5.4). It is thus twice as likely to roll a seven as it is to roll a four, and *six* times as likely

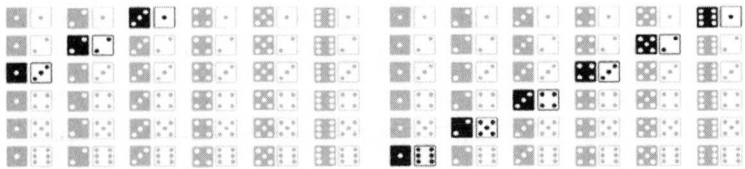

Figure 5.4

Out of thirty-six possible rolls of two dice, there are six different ways to roll two dice to total seven (*right*), but only three ways to roll dice that total four (*left*). Since any roll is equally likely, this means rolling four is half as likely as rolling seven.

as to roll a two or twelve. Figure 5.5 shows the result of a hundred thousand rolls of two dice; the most likely outcome (one-sixth of the rolls) is a seven, with other rolls progressively less likely for numbers further away from seven. The result is a distribution of outcomes that approximates the bell-shaped curve of an exact Gaussian. (This example is only approximate; a roll of two dice forms a *binomial* distribution, which is not exactly Gaussian. To achieve an exact Gaussian requires one to roll an *infinite* number of dice!)

Similarly, the amplitudes of vacuum quantum modes produced during inflation are selected from a Gaussian statistical distribution. Since the specific form of the perturbations is by definition random, it is impossible to calculate, just as it is impossible to predict the outcome of a single roll of two dice. It is, however, possible to describe the primordial perturbations in a statistical sense, just as it is possible to predict the distribution of a large number of rolls of the dice (figure 5.5). Inflation

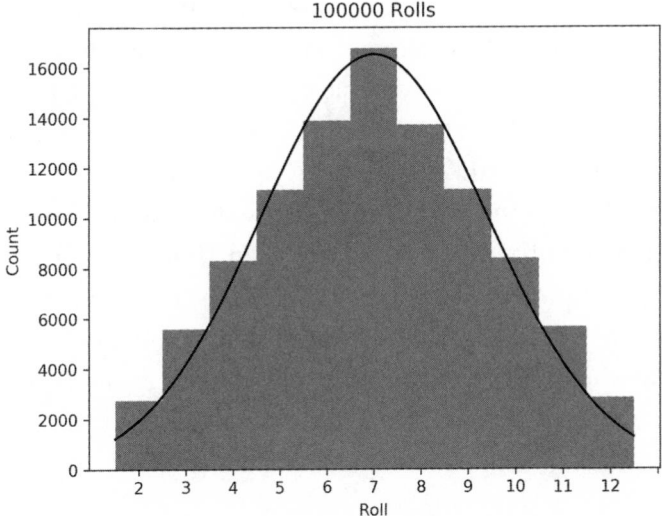

Figure 5.5
The results of a hundred thousand rolls of two dice, showing the resulting count for each possible outcome. The black curve depicts the corresponding Gaussian probability distribution.

also makes a very definite prediction about the relative amplitudes of the perturbations at *different* wavelengths. Quantum modes that are stretched outside the horizon earlier in the inflationary epoch are stretched more than quantum modes that exit later on, closer to the end of inflation. Hence modes that exit the horizon earlier in the inflationary epoch have a longer wavelength at the end of inflation than modes that exit later. Because the expansion rate during slow roll changes only very slowly, very long and very short wavelength perturbations should on average be almost exactly the same—a property known as *scale invariance*.

In fact, inflation does not predict perfect scale invariance because the expansion rate during inflation is not constant but instead very slowly decreasing as the field slowly rolls down the potential surface. This means that long wavelength perturbations (which exit earlier in inflation, with a higher expansion rate) will on average be a slightly higher amplitude than short wavelength perturbations, which exit later when the expansion rate is slightly lower. This is expressed mathematically by the *power spectrum*, a measure of the average amplitude of perturbations as a function of the wavelength λ:

$$P \propto \lambda^{1-n}$$

This power-law form is a prediction of slow roll inflation, where the exponent n is called the *spectral index*. A scale-invariant distribution, with equal power at all wavelengths, has spectral index $n = 1$. A distribution with more power on longer wavelengths and less on shorter ones has $n < 1$. This is called a *red* spectrum. *All* inflation models predict a slightly red power spectrum for gravitational waves. This comes from the fact that the gravitational wave modes depend on the expansion rate during inflation, which must always decrease with time. Single-field potentials of the type pictured in figures 4.5 and 4.7 likewise predict a red power spectrum for density perturbations. Since the density perturbations depend on both the expansion rate and rate at which the field rolls down the potential, it is possible to engineer models for which density perturbations have more power on small wavelengths (called a *blue* spectrum, with $n > 1$), but we will see that such choices of potential are disfavored by data.

I am now in a position to discuss the predictions of inflation in light of cosmological observations, which is the subject of the next chapter.

6
Testing Inflation

If it doesn't agree with experiment, it's wrong.

—Richard Feynman

The Big Picture

Any useful scientific theory must make predictions that can be tested against observations of nature. We have seen that cosmological inflation—originally motivated as a way to explain the large-scale flatness and homogeneity of the universe, and absence of primordial relics such as magnetic monopoles—predicts the stretching of quantum zero-point modes as an *inevitable* consequence of accelerating expansion. It is these quantum modes that generate the primordial fluctuations in the density of the cosmos and create the initial conditions for cosmic structure formation. This remarkably powerful synthesis of gravity and quantum mechanics enables inflation to provide a single, unified picture for the physics of the very early universe as well as the initial conditions for the hot big bang. Furthermore, the quantum nature of the primordial perturbations generated by inflation means that their predicted form can be precisely specified. Slow roll inflation makes a set of definite, testable predictions for the form of the deviation from perfect homogeneity in the universe:

- Inflation predicts the generation of both density perturbations (also called *scalar* perturbations) and primordial gravitational waves (also called *tensor* perturbations)

- Both scalar and tensor perturbations should be nearly but not exactly scale invariant, with a power-law dependence of amplitude on wavelength

- Both scalar and tensor perturbations should be generated with Gaussian random statistics

- The perturbations created during inflation will be *superhorizon*—that is, have wavelengths larger than the horizon size of the universe

These specific properties are a consequence of the quantum origin of the primordial perturbations in inflation. (Other mechanisms for generating the seed perturbations for structure in the universe—for example, *topological defects* such as cosmic strings—make very different predictions.) This means that the inflationary paradigm can be tested by measuring the form of the primordial density perturbations in the universe. This is one of the key goals of twenty-first-century cosmology and the subject of intense activity since these primordial perturbations form the initial conditions for structure formation too. By far the most powerful tool right now is the CMB, which has been measured with exquisite precision by the National Aeronautics and Space Administration's WMAP satellite and the European Space Agency's follow-up Planck mission as well as a wide array of ground- and balloon-based experiments.[1] Data from the CMB are complemented by measurements of the three-dimensional distribution of galaxies in the universe by the Sloan Digital Sky Survey, which released its first data in 2003, and has continued through the SDSS-III/BOSS Survey.[2]

These observations, taken together, result in a ringing success for the theory of inflation, which I summarize in this chapter. Inflation makes a number of specific, highly nontrivial predictions for the primordial universe. Every one of those predictions has been validated by data, with the exception of primordial gravitational waves. (I will talk more about this last case below.) This is the big picture: inflation makes real, testable predictions, and those predictions have been tested—and confirmed—with high-precision cosmological observations. But there is more to the story. The era of precision cosmology allows us to ask new questions about inflation—most important, to distinguish one inflationary model from another.

Let me take the general predictions of inflation one at a time, starting with superhorizon perturbations.

Superhorizon Perturbations

By far the most powerful general test of inflation is the presence of superhorizon perturbations. This is a prediction that is, at least in the case of an expanding universe, *unique* to inflation (I will explore alternatives in chapter 8), and does not depend on any particular model for inflation. No purely causal theory in an expanding universe can produce superhorizon perturbations. Searching for superhorizon modes in the CMB is in principle a simple exercise. We saw in chapter 3 that the horizon size at the time of recombination corresponds to an angular scale on the sky today of about a degree. Therefore a signature of superhorizon perturbations is statistical correlations in the CMB anisotropy on scales *larger* than a degree. In the case of the temperature anisotropy shown in figures 3.6–3.7, this is complicated by the presence of structure in the current universe, which can create "foreground" signals that mimic primordial correlations. For this reason, the most sensitive current probe of the presence of primordial superhorizon perturbations is the *polarization* of the CMB, which is less contaminated by signals from late-time structure.

The cleanest signal for superhorizon perturbations is in the statistical correlation between the polarization and fluctuations in the temperature of the CMB, called the *cross-correlation* of the temperature and polarization. Figure 6.1 shows the cross-correlation as measured by the Planck satellite.[3] As in figure 3.7, the cross-correlation is

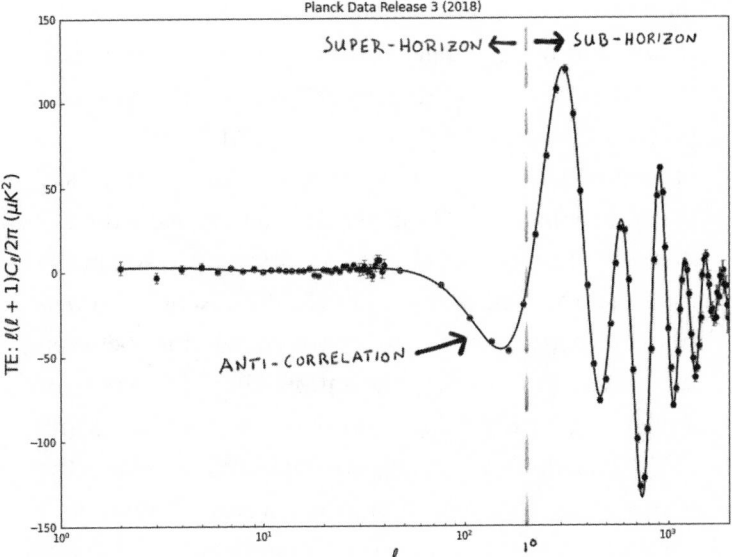

Figure 6.1

The smoking gun signal for superhorizon cosmological perturbations in the correlation between temperature and polarization as measured by the Planck satellite (*dots*). The best-fit LCDM cosmology is shown by a solid line. Large angles on the sky are to the left, and small angles are to the right. The dashed vertical line is at one degree, the size of the horizon at recombination. The data show clear anticorrelation on angular scales larger than one degree—a signature of superhorizon perturbations from inflation. *Sources*: Data: ESA/Planck collaboration; plot by the author.

plotted as a function of angular scale, with larger angular scales (and hence longer wavelength perturbations) on the left, and smaller angular scales (shorter wavelength perturbations) on the right. The horizon size at the time of recombination subtends about one degree on the sky today, so angular scales larger than a degree on the sky measure distances larger than the cosmic horizon at the time that the CMB was emitted. In the absence of super-horizon perturbations, the cross-correlation between the temperature and polarization should vanish for angles larger than a degree in size. What we see clearly in the Planck data is a negative correlation between the temperature and polarization on scales larger than a degree on the sky, which is an unambiguous signal of superhorizon density perturbation modes. Here *anti-correlation* means that larger temperature fluctuations are statistically associated with negative polarization on very large angular scales. This smoking gun signal cannot be created by structure in the local universe via any known process.

The second smoking gun for inflation is primordial gravitational waves, which I discuss in the next section.

Primordial Gravitational Waves

The second key signature of inflation is the presence of primordial gravitational waves. Such gravitational waves are not within the sensitivity of current direct detection experiments such as the LIGO/Virgo detector, but they can be searched for via indirect effects, particularly the polarization of light from the CMB. Both density perturbations and gravitational waves cause the CMB light to become polarized, but the patterns of polarization created by the two sources are distinctly different. Density perturbations polarize the CMB light with either a circular pattern (created by cold spots in the CMB) or radial pattern (created by hot spots), as shown in figure 6.2. This *E-mode* polarization was first detected by the DASI telescope in 2002, and has subsequently been measured to a high precision by experiments such as the Planck satellite.[4] Primordial gravitational waves (also called *tensor* perturbations) produce E-mode polarization as well, but they generate polarization with a type of pattern that density perturbations do not. This second type of polarization, called *B-mode*, is helical, winding either clockwise or counterclockwise (figure 6.3).[5] While such a pattern of polarization in the primordial CMB is unique to gravitational waves, it can be created by late-time astrophysics too, especially the gravitational lensing of the E-mode polarization by cosmic structure, and by emission from dust inside our own galaxy. The observational challenges in detecting a primordial B-mode are daunting; the primordial B-mode is in fact expected to be smaller than these "foreground" signals, which must be removed. The lensing and dust B-mode signals have been well measured, but as of yet there has been no detection of a *primordial* B-mode and thus no

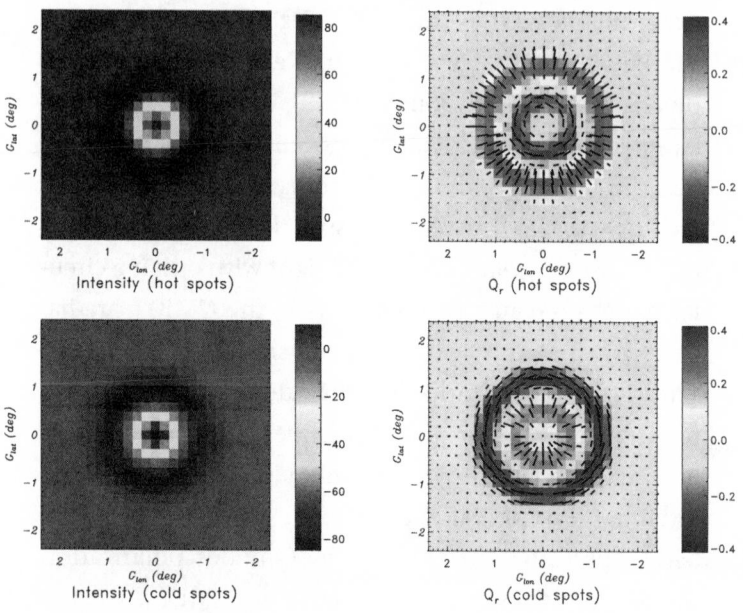

Figure 6.2
E-mode polarization observed by the Planck satellite, shown as lines superimposed on the temperature anisotropy. The polarization is radial in cold spots and circular in hot spots. *Source*: ESA/Planck collaboration.

detection of the primordial gravitational waves predicted by inflation.[6]

Current bounds on primordial gravitational waves are not, however, the end of the story. We cannot unambiguously rule out the presence of gravitational waves from inflation: instead, the best we can do is place an upper bound on their amplitude, and inflation makes no particular prediction for this. It is entirely possible that primordial gravitational waves from inflation are

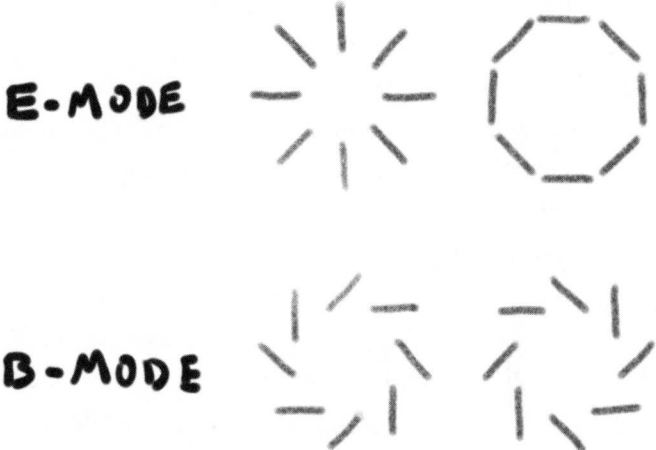

Figure 6.3

E-mode and B-mode polarization patterns. B-mode polarization has a "handedness," and E-mode polarization does not.

there, but are too weak to be detected by current measurements. Future high-precision CMB measurements promise to significantly improve the bound on primordial gravitational waves, which I discuss later in this chapter. In the meantime, even an upper bound, combined with measurement of the weak scale dependence of density perturbations, gives us a powerful probe of inflationary physics, which I turn to next.

Scale Invariance: The Zoo Plot

We have seen that the most general predictions of cosmological inflation have been spectacularly confirmed by modern precision cosmological data—most significantly the presence of superhorizon perturbations in the early universe. A second prediction, the presence of a background of primordial gravitational waves, has yet to be confirmed, but searches are ongoing. Precision data allow us to do much more and begin to ask questions about the physics of inflation itself. What are the properties of the field (or fields) responsible for inflation? At what energy scale does inflation take place? What is the form of the inflationary potential? Is the dynamics of inflation governed by one field or more than one? Because primordial perturbations are created from quantum zero-point modes in inflation, the form of the perturbations give us a direct probe of the physics of inflation itself. The key to this effort is the relationship between the primordial perturbations and shape of the potential surface for inflation.

Primordial gravitational waves (*tensor perturbations*) and primordial density perturbations (*scalar perturbations*) differ in key ways. The amplitude of the tensor perturbations depends only on the expansion rate during inflation, which itself is determined by the energy scale of inflation. The higher the energy density during inflation, the higher the expansion rate, and the stronger the gravitational waves produced. Scalar perturbations depend on both the expansion rate and the rate at which the field "rolls" down its potential. If we can measure both scalar and tensor perturbations, it becomes possible to disentangle these effects and gain information about not only the energy scale of inflation but also the *shape* of the potential surface associated with symmetry breaking in the very early universe. This is an exciting possibility.

To make the connection between cosmological data and fundamental properties of the inflaton field, we must translate between the language of quantum field theory into that of observable astrophysics. Since both scalar and tensor perturbations contribute to the temperature fluctuations in the CMB, a convenient parameter is the tensor fraction (or *tensor/scalar ratio*) r, which is defined as the fraction of the observed perturbations created by tensor (gravitational wave) modes. That is, $r = 0.1$ means that 10 percent of the perturbations we observe in the CMB are due to tensor modes, and $r = 0.01$ means that tensors contribute 1 percent. This number is useful because r can be translated directly into the energy scale Λ of inflation, in units of billions of electron volts (GeV):

$$r = \left(\frac{\Lambda}{3.3 \times 10^{16} \text{ GeV}} \right)^4.$$

Two things are especially notable about this relation. The first is the enormous energy scales involved. The fission of a uranium atom releases about 200 MeV of energy, and collisions at the LHC have an energy of about 10 trillion electron volts, or 10,000 GeV. A tensor fraction of the order of a few percent corresponds to an energy scale of around 10^{15} billion electron volts, or a *100 billion* times the energy of the LHC. This is a remarkable meeting point of the physics of outer and inner space; we can use the universe as a natural particle accelerator to probe energy scales vastly beyond the reach of human-constructed instruments. This is the good news. The bad news is that the ratio r varies as the *fourth power* of the energy scale of inflation. This means that if we lower the energy scale of inflation by a factor of 10, the tensor fraction drops by a factor of *10,000*. A small shift in the energy scale of inflation moves the gravitational waves from the early universe from easily observable to completely out of reach of feasible future measurements. It is perhaps not surprising that primordial tensors have yet to be detected, but we will see that even an upper bound on the tensor fraction is still a useful probe of the physics of inflation.

Each particular choice of potential makes a different prediction for the tensor fraction r and spectral index n of the scalar perturbation amplitude, measuring the deviation from scale invariance ($n = 1$). For a given inflation model, these can be precisely calculated via quantum field theory. The form of the potential surface during inflation can be translated into precise predictions, which can then be compared to observation. This can be represented graphically by a *zoo plot* of tensor fraction r versus spectral index n.[7] Figure 6.4 shows a zoo plot including a few choices of inflationary potential, and bounds from the Planck satellite combined with the upper bound on the B-mode polarization of the CMB

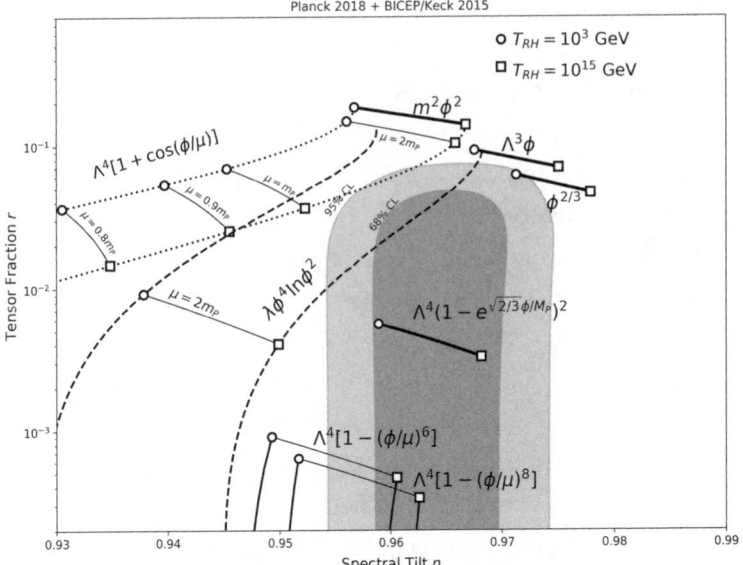

Planck 2018 + BICEP/Keck 2015

Figure 6.4

A zoo plot of tensor fraction r versus scalar spectral index n. The lines represent the predictions of different choices of potential surface for inflation, and the shaded regions show the regions allowed by Planck and BICEP/Keck at a 68 percent (*dark shaded*) and 95 percent (*light shaded*) statistical confidence. (Figure by the author.) T_{RH} denotes the reheat temperature at the end of inflation, which influences the predictions for each choice of potential.

obtained with the BICEP/Keck telescope at the South Pole.[8] Both r and n are measured to high precision, with an upper bound on the tensor fraction $r < 0.07$, and scalar spectral index n between 0.955 and 0.974 (with 95 percent statistical confidence), shown in the shaded region. The different inflationary potentials are labeled by their form as a function of the field ϕ. Note that most of the potentials plotted—for example, $V(\phi) = m^2\phi^2$, land outside the region allowed by the CMB data. That means that these choices of potential are ruled out to a high statistical confidence. This is transformative. Precision cosmology has reached the point where it is not only possible to test the broad predictions of inflationary theory but also possible to test *specific models* for inflation and therefore test the underlying particle physics, at energy scales vastly beyond the reach of terrestrial accelerators. Existing bounds are crude, but it is remarkable that any bound is possible at all.

What do we learn about the physics of inflation? One thing is that simple, single-field models of inflation of the sort I looked at in chapter 4 are completely consistent with the data. But we can go further and ask *which* single-field models fit, and which do not. Are the data consistent with a symmetry breaking potential of the type sketched in figure 4.5 or a potential with no breaking of symmetry as portrayed in figure 4.7? Figure 6.5 shows the result. The different types of potential occupy distinct regions of the zoo plot, with concave-up potentials of the type pictured in figure 4.7 on the upper right, and symmetry breaking potentials (figure 4.5) on the lower left. The data clearly favor the symmetry breaking potential, with the field rolling off an unstable hilltop. The potential that best fits the data is the type depicted in figure 4.9, characterized by a long, flat "plateau" enabling slow roll. This is shown quantitatively in figure 6.4 as the potential labeled $\Lambda^4\left(1 - e^{\sqrt{2/3}\,\phi/M_P}\right)^2$, which was proposed by Russian physicist Starobinsky in

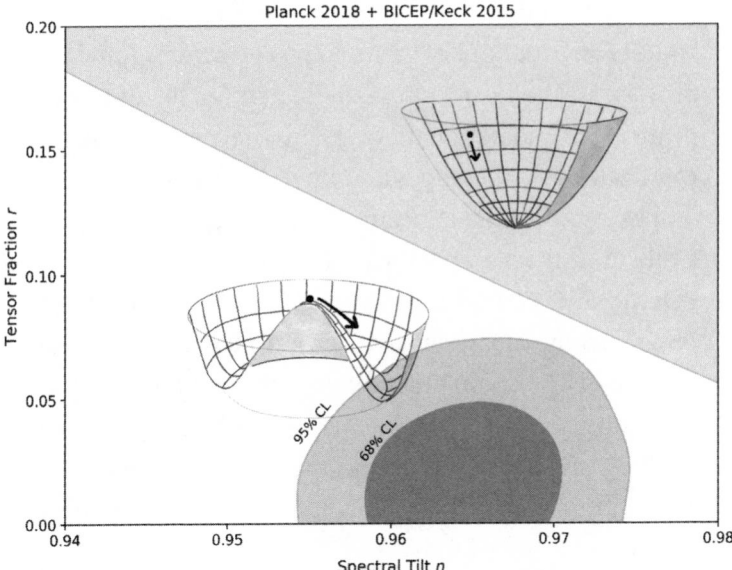

Figure 6.5
Concave-up potentials with no symmetry breaking and concave-down potentials typical of symmetry breaking occupy different regions of the zoo plot. The current data favor concave-down potential surfaces.

1980.[9] It is ironic, given the thousands of papers written over the course of decades constructing possible models for inflation, that one of the earliest and simplest models turns out to be the best fit to the data.

Future Tests: Consistency Relations

The story of precision cosmological tests of inflation is by no means over. New experiments being developed promise to substantially improve the precision of constraints on inflationary cosmology. The most significant of these in the near term is the search for signatures of primordial gravitational waves in the polarization of the CMB. New telescopes and upgrades to existing telescopes such as the Atacama Cosmology Telescope in Chile, South Pole and BICEP telescopes in Antarctica, POLARBEAR-2/Simons Array, and QUIET will construct progressively more sensitive ground-based observatories. These are complemented by the high-altitude balloon observatories SPIDER and EBEX. The proposed CMB S4 (for "Stage 4") project aims to conduct measurements capable of unambiguously detecting a tensor fraction of order $r = 0.01$.

Inflation involving a single field makes additional, highly specific predictions that (at least in principle) could be subject to future tests, known as *consistency conditions*. So far I have discussed in detail the spectral index n, which describes the scale dependence of the power-law density perturbation spectrum. Yet the gravitational wave power spectrum is also predicted to be almost (but not quite) scale invariant, and its scale dependence is described by a corresponding spectral index n_T. In single-field inflation, however, the tensor spectral index is *not* an independent parameter but instead directly related to the tensor/scalar ratio r,

$$n_T = -\frac{r}{8}.$$

If inflation involves more than one field or if inflation itself is wrong, this relation will not hold. As such, the single-field consistency relation represents a compelling target for future observational tests. This is easier said than done, though, since (as we have already seen) the tensor spectrum may well be too small to be detectable, at least in the near future. Furthermore, testing this consistency relation requires not just *detection* of primordial gravitational waves but detection with sufficient accuracy to measure the dependence of the spectrum on wavelength too. This is a considerably higher bar; in the case of scalar perturbations, the time between the first detection by COBE and first reliable measurement of the spectral index by WMAP was a decade. In the case of tensor modes, the feasibility of such a measurement depends crucially on the value of the tensor fraction, which is currently unknown. Precision tests of the single-field consistency relation will most likely involve indirect detection via CMB polarization, followed by direct detection of primordial gravitational waves at short wavelength—for example, by a space-based gravitational wave detector such as the proposed Big Bang Observer (the United States) or DECIGO (Japan).[10]

The second future frontier for tests of inflation is the measurement of deviations from perfectly Gaussian random statistics of the primordial perturbations. We saw in chapter 5 that the quantum nature of perturbations in inflation leads to Gaussianity, but this is a particular and testable prediction. To think about measuring *non*-Gaussianity, think about our analogy of rolls of dice in chapter 5. Suppose the dice were loaded—that is, not perfectly random. If you were a player at a casino, how would you tell? One way to do so would be to look at a large number of rolls and plot the distribution of outcomes, as in figure 5.5. If you observed, for example, more outcomes at the extremes—rolls of two or twelve—than predicted by chance, it would be a good sign that the dice were not fair. One can do exactly the same thing with cosmological perturbations. One instance of an extreme outcome is very large perturbations in the density, such as those seeding clusters of galaxies. An excess of such extreme perturbations would be reflected in an anomalously large population of large galaxy clusters, which would be evident in cosmological galaxy surveys. Similarly, too many large perturbations in the CMB temperature—or too few—would be a signal for non-Gaussian statistics.

Inflation, in fact, does not predict *perfect* Gaussianity. A consistency relation, first derived by Juan Maldacena in 2003, relates the spectral index n to the deviation of the fluctuations from perfectly Gaussian random statistics.[11] To the lowest approximation, single-field inflation predicts perfectly Gaussian random perturbations, but a more careful analysis shows that there should in fact be tiny *non-Gaussianity* caused by coupling between very long and very short wavelength perturbations. This coupling arises because superhorizon perturbations rescale the density of the local horizon volume.

This results in a change in the expansion rate over the locally observable patch universe. If a particular horizon volume is slightly more dense than the average over larger than the horizon size, it will appear to be expanding slightly more quickly to an observer *inside* the horizon. Likewise, a horizon that is slightly less dense will appear locally to be expanding slightly more slowly. Long wavelength perturbations exit the horizon first and modify the local expansion rate when shorter wavelength perturbations exit the horizon later on. These slight mismatches in the expansion rate from one location to another influence the amplitude of the short wavelength modes. This modulation of short wavelength perturbations by long ones is referred to as *local* non-Gaussianity and is shown schematically in figure 6.6. Maldacena showed that single-field, slow roll inflation makes a definite prediction about the local non-Gaussianity, measured by a dimensionless number f. The amount of local non-Gaussianity in such cases is proportional to the departure from scale invariance in the scalar perturbations,

$$f = \frac{5}{12}(n-1).$$

Like the consistency relation connecting the tensor spectral index n_T and tensor fraction r, the consistency relation between the local non-Gaussianity f and scalar spectral index n is a specific prediction of single-field, slow roll inflation. The current best limit on the local

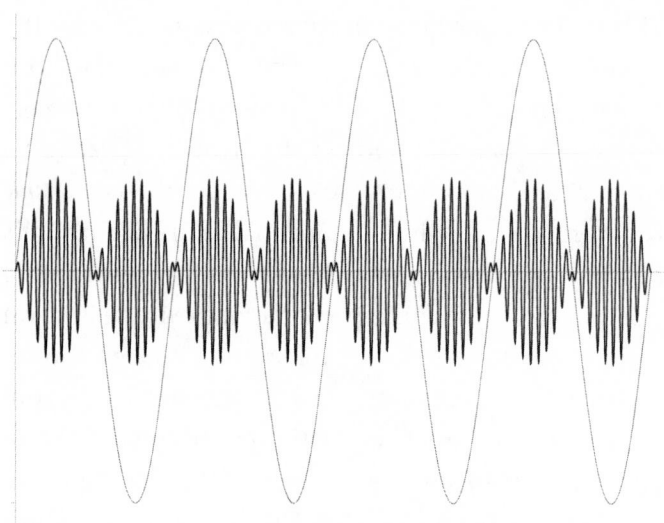

Figure 6.6

A schematic representation of local non-Gaussianity. A short wavelength mode (*solid line*) is modulated by a long wavelength mode (*dashed line*). Coupling between long and short wavelength modes introduces non-Gaussianity into the primordial perturbations.

non-Gaussianity comes from the Planck satellite, which did not detect non-Gaussianity, but set an upper bound of $f < 4.2$.[12] Yet the Planck best fit for the scalar spectral index is $n - 1 = 0.04$, which means that the predicted non-Gaussianity from Maldacena's consistency relation is $f = 0.02$, or about two hundred times smaller than the Planck bound. This bound will be improved by a factor of a few by upcoming measurements of cosmic large-scale structure.[13] Achieving the precision necessary to test the inflationary consistency condition is a substantial observational challenge, involving proposed radio telescope arrays operating on the far side of the moon.[14] Conversely, a detection of local non-Gaussianity of $f > 0.1$ would conclusively rule out *all* single-field, slow roll inflation models such as those plotted in figure 6.4.

Long wavelength perturbations create an additional effect besides local non-Gaussianity: curvature. We saw in chapter 4 that inflationary evolution drives the universe to geometric flatness, but this is a purely classical effect. When we include quantum mechanics, this picture must be modified. Quantum vacuum fluctuations during inflation generate perturbations in the density at all scales, both short and long wavelengths. Wavelengths shorter than our cosmic horizon today form the seeds for structure formation, but wavelengths much longer than our horizon still have an effect, appearing as geometric curvature. Classically, inflation drives the curvature exponentially close to zero, but quantum mechanics puts a lower bound on the curvature. Like the density perturbations that form structure, this lower bound is expected to be of order 10^{-5}, or about three orders of magnitude smaller than the upper bound on the curvature set by current observations of the CMB. This model-independent prediction is an achievable target for future observations; a failure to measure a curvature of order 10^{-5} would cast doubt on the viability of inflation in general.

Variations on a Theme

Inflation generated by a single scalar field, such as the simple case of a phase transition as discussed in chapter 4, is a scenario that is completely consistent with all present data. Most important, such models are specifically *predictive* scientific theories. The simplest predictions, such as near scale invariance, superhorizon correlations, and Gaussian perturbations, have already been confirmed to a high precision—a nontrivial achievement. The universe could have turned out quite differently, ruling out such models altogether. Furthermore, single-field inflation makes predictions that have not yet been tested: primordial gravitational waves, curvature, and consistency relations for the tensor spectrum and local non-Gaussianity. I hope I have convinced you in this chapter that inflation is a well-posed scientific theory in the classic sense of predictivity and falsifiability—in fact with great detail and precision.

Even within the specific framework of single-field inflation, however, particular models based in various theories of high-energy physics such as supersymmetry and string theory proliferate. Theorists are a busy lot, and over the last couple of decades, literally thousands of particular models have been proposed, involving a dizzying array of possibilities for the fundamental nature of the field driving inflation. The key point is that the shape of the potential surface on which the inflationary field evolves is a free function—that

is, it can be *any* shape, as long as it is sufficiently flat
to support accelerated expansion. This leaves considerable freedom; realizations of inflation in different
fundamental theories will pick out different particular
shapes for the potential surface—for example, because
of specific symmetries in the theories—but there is no
a priori constraint on the shape of the potential. This is
a question that must ultimately be resolved by observation; for instance, the current data prefer a hilltop-type
potential over a simple bowl shape (figure 6.5). One
interesting question one might ask is whether or not
there are "preferred" values of observables such as the
tensor fraction, which are in some sense more or less
likely to occur in single-field inflation. Numerical tech-
niques prove useful here, particularly the *flow formalism*
initially proposed by Mark Hoffman and Michael Turner
in 2001, and later expanded by me in 2002.[15] The flow
formalism makes it possible to generate large numbers
of different inflationary universes in a computer instead
of hand designing potential surfaces one at a time. This
means that we can study the *generic* predictions of inflation rather than relying on particular models for high-
energy physics. Figure 6.7 shows the results of such a
numerical approach, plotting one million inflation
models (*points*) over the parameter regions favored by
the Planck CMB data (*shaded region*). The lesson learned
is that the points densely fill the plot, indicating that
there is no particular value of tensor fraction or spectral
tilt that is favored in single-field inflation. This inherent

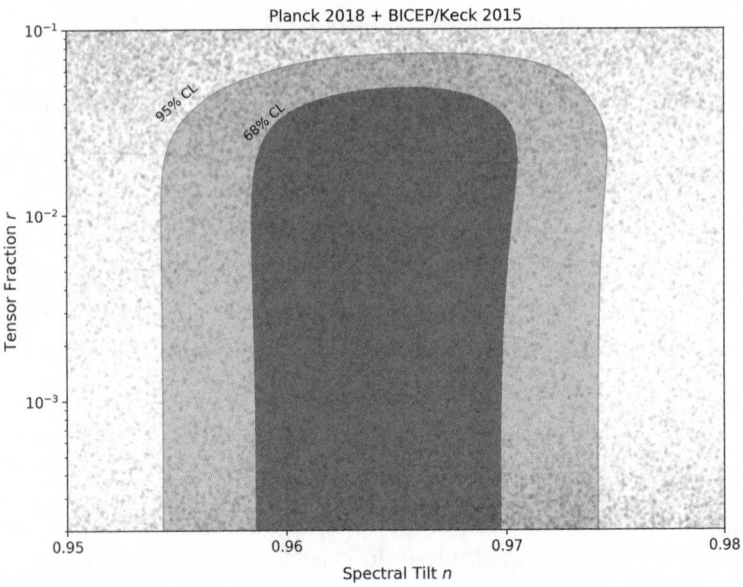

Figure 6.7
One million single-field inflation models (*points*) plotted by tensor fraction and spectral tilt. The shaded region shows the parameters allowed at 68 percent (*inner*) and 95 percent (*outer*) confidence by the Planck CMB measurement, combined with the BICEP/Keck measurement of CMB polarization. The models densely fill the parameter space, suggesting that there is no region favored by single-field inflation.

model dependence is sometimes confused with a lack of predictivity; if inflation can predict any outcome, is it really a scientific theory at all, or simply a "just so" story, which can be adjusted at will to fit any data at all? I will return to this question in chapter 8, but one of the purposes of the discussion here is to demonstrate that this is a misguided criticism. Single-field inflation, for example, may be right or wrong as a description of the early universe, but it makes a definite—and testable—set of predictions about observations, and (so far) these predictions have been beautifully confirmed. Within those constraints, inflation contains significant freedom to fit specific parameters, such as the tensor fraction, depending on the underlying particle physics model. This is no different than any other physical theory; Newton's law of gravity, for instance, explains how the planets orbit about the sun, but it does not predict the number of planets in the solar system or their orbital radii.

Inflation involving a single field slowly rolling along a potential surface is not the only possibility, and in fact the space of models is far wider. Possible extensions include:

- *Multifield inflation*: Such models involve multiple fields directly in the dynamics driving inflation—a considerably more complex case than I have explored here. Multifield models can produce distinctive observational signatures, such as large local non-Gaussianity or a violation of the consistency relation of the tensor fraction to the spectral tilt.

- *Noncanonical inflation*: The single-field models I have looked at far themselves form a restricted set of physical theories, referred to as *canonical* field theories, for which the speed of sound in the field is equal to the speed of light. Theories of quantum gravity such as string theory allow for a larger class of field theory for which the speed of sound is less than (or sometimes even *greater than*) the speed of light. Theories with a low speed of sound create large non-Gaussianities in the fluctuations produced during inflation. These take a form distinct from the local non-Gaussianity produced in multi-field models, called *equilateral* non-Gaussianity. No non-Gaussian features of the equilateral type were detected by the Planck CMB measurement, placing a lower bound on the speed of sound c_S during inflation of 8 percent of the speed of light.

- *Warm inflation*: In "warm" models, space during inflation is not empty but instead filled with a thermal-equilibrium background of particles that does not dilute during inflation. Friction on the inflationary field from the thermal bath dominates over friction from the cosmic expansion, which means that inflation can take place on potential surfaces too steep to support slow roll inflation.

- *Modified gravity theories*: The dynamics of inflation can be significantly altered in alternatives or extensions to Einstein's general theory of relativity. One particularly tantalizing possibility is that exotic couplings to gravity can make it possible that the Higgs field from the standard model—a known particle—could itself be responsible for inflation.[16]

- *Spectator theories*: In simple single-field inflation, the inflaton field is responsible for both the accelerating expansion and generation of density perturbations in the universe. In more general scenarios, these roles can be split, with the inflaton governing the background expansion, but the perturbations generated by another field, along for the ride during inflation. This second field is called a *curvaton*. Current data disfavor curvaton scenarios, which typically produce too large a contribution to the CMB from local non-Gaussianity.

While such variations are *possible* and represent unique opportunities for interesting model-building, none of them are *necessary* in the sense of being required by observation. The simplest single-field, slowly rolling inflation models are entirely sufficient to explain the existing data. This of course may change when future high-precision measurements become available, but applying Occam's razor—the simplest explanation is more likely to be the right one—leads one to favor single-field models over the more complex possibilities, all else being equal.

This is a remarkable success story. Starting as a theoretically motivated argument explaining the large-scale properties of flatness and homogeneity in cosmology, inflation has been developed over the period of a few decades into a precise and predictive scientific theory. It is a tour de force of twenty-first-century cosmology that predictions of inflation have indeed been tested and confirmed—a feat that would have been barely thinkable in the early 1980s when the theory was first proposed. My memorable first introduction to the theory was as an undergraduate student, speaking with the cosmologist Edward Groth in his office in Jadwin Hall on the campus of Princeton University in spring 1985. "What," he asked me, entirely out of the blue, "do you think of this new *inflation* theory?" I confessed that I had never heard of it, and he replied, in his typically irascible fashion, "Well I think it's *bullshit*." And that was that. It was perhaps overly hasty, viewed in retrospect, but the famous cosmologist's judgment—even in light of the experimental success of the theory—contains wisdom. It turns out that the situation is not as simple as it may seem. Any good scientific theory must not only provide answers to our existing questions about the world but should also suggest new questions to pursue—questions that we could not have asked before. In this spirit, inflation as a successful paradigm raises new questions about the global structure of space and time, the most profound of which fundamentally change our picture of our place in the cosmos and force us to rethink what it is possible to *know* about the universe at all. I take this up in the next chapter.

7
Eternal Inflation and the Multiverse

Chance governs all. Into this wilde Abyss,
The Womb of nature and perhaps her Grave,
Of neither Sea, nor Shore, nor Air, nor Fire,
But all these in thir pregnant causes mixt
Confus'dly, and which thus must ever fight,
Unless th' Almighty Maker them ordain
His dark materials to create more Worlds
                              —John Milton, *Paradise Lost*

Ptolemy's universe was one of order. Informed by Aristotle's formulation of physics, in which the natural state of a material body is at rest, Ptolemy placed the Earth at the center of the universe—a "bottom" at which all things material collect. Ptolemy himself was not Christian, but his model was adopted enthusiastically by the Roman Catholic Church, particularly in the writings of Thomas Aquinas. In Aquinas's vision, above the Earth is the realm of the celestial spheres, perfect, inhabited by God and the angels. Below the earth is hell, to which all things sullied and evil sink. In between are humans, material creatures in an imperfect world, torn between divinity and sin. Our position in the cosmos makes us special creatures, made in the image of God, and capable of transcending the imperfection of Earth and achieving the perfection of heaven. The specialness of the Earth, and by extension of us, is not an accident. In Ptolemy's universe, our distinctive place is built into the structure of the universe itself, or as Aquinas put it, "God-ordained and man-centered." The Earth is singular, the sum total of all that is material, since the laws of physics require that all material things fall to rest at the center of the cosmos. Ptolemy's universe is inherently hierarchical too. Ethereal heaven and material Earth are separate, and governed by different laws, and Earth's special nature is not one of privilege but rather inferiority. God, in the perfect realm of heaven, rules the imperfect Earth. It is perhaps no accident that the social and political structures of the Middle Ages reflected the hierarchical cosmology of Ptolemy. The dominion of the church and divine right of kings were mirrors of the structure of the cosmos itself. Likewise, Aquinas's formulation of ethics was fundamentally based on his conception of the cosmological order: "The judgment, however, of the goodness of anything does not depend upon its order to any particular thing, but rather upon what it is in itself, and on its order to the whole universe, wherein every part has its own perfectly ordered place."[1]

Copernican ideas challenged that order. In his 1584 work *On the Infinite Universe and Worlds*, the Copernican scientist and mystic Giordano Bruno proposed that the greatness of God must inevitably be reflected in an infinity of worlds, writing that "God is infinite, so His universe must be too. . . . He is glorified not in one, but in countless suns; not in a single earth, a single world, but in a thousand thousand, I say in an infinity of worlds."[2] It is hard to imagine coming from a modern perspective how such a pious argument could have been considered so heretical and dangerous. But if the social and political order in the material world derives its legitimacy from the inherent order of the universe itself, then the idea of an infinity of worlds utterly destroys that order, and by extension the very foundation of medieval society. Copernicanism in general and Bruno's ideas in particular were deeply radical. Bruno was imprisoned for seven years and ultimately burned at the stake at Campo di Fiori for heresy by the Inquisition in 1600. Two of the eight counts of heresy at Bruno's trial were his doctrine of an infinity of worlds and the doctrine of terrestrial movement. At his sentencing, understanding the situation perfectly, he told the judges, "Perchance you who pronounce my sentence are in greater fear than I who receive it."[3] The world was changing.

If Bruno was right and we are one of an infinity of worlds, we are not special since in an infinity of worlds, there will certainly be an infinity of civilizations, even an infinity of other Earths, identical to our own. In a truly infinite cosmos, we are as individuals not even unique manifestations of ourselves but instead one of an infinity of copies, which changes our picture of ourselves in basic ways. Bruno's infinity of worlds has a spectacular realization in modern cosmology in the idea of eternal inflation.

Eternal Inflation

We have seen that the two basic features underlying the physics of inflation are:

1. Size doubling
2. Instability

The first is necessary for inflation to stretch a tiny initial patch of whatever was the preinflationary state to a macroscopic size, large enough to encompass the universe that we live in and observe today. The second feature, instability, is necessary because inflation, if it is to match the universe we see, needs to *end* and transition to a hot, thermal equilibrium state during the period of reheating. The end of inflation is in fact all we see; anything that happened before the final eighty doublings in size of the universe—about a factor of 10^{24} in size—is larger than our current cosmological horizon and is unobservable. This raises an obvious question: How long can this period of inflation go on? Is there an upper limit to the number of doublings in size allowed by basic principles of physics? The answer here is not so clear. We must have at *least* eighty doublings, but if there are a hundred, a thousand, or ten thousand this will have no effect on the properties of *our* observable universe. (One might also ask, If there are no observable consequences of these additional size doublings, does the question have any scientific meaning? I will return to this question later.) To understand the physical principles at play, I must return to the discussion of the physics of instability.

Consider again the pencil balanced on its tip. Newton's laws of motion tell us that a pencil, perfectly balanced at the vertical, will remain vertical *forever* unless some outside force acts on it. Pencils in the real world fall over because of external forces such as air currents, but even if we were to remove all such outside forces and perfectly isolate our precisely balanced pencil from its surroundings, we would have quantum mechanics to contend with. The Heisenberg uncertainty principle tells us that it is impossible to localize the pencil precisely at both the vertical and at rest. Tiny quantum fluctuations will always pull the pencil away from its perfect equilibrium state, just enough to initiate its inevitable fall. Quantum uncertainty ensures that unstable equilibriums, even in perfect isolation, have finite lifetimes. In the case of unstable equilibriums in inflation, there is an additional element: expansion. Expansion changes things in two ways. The first is expansion drag, which slows the evolution away from equilibrium. The second is the universe doubling in size, and doubling again, and again as inflation progresses. Size doubling has no analog in simple dynamical systems like pencils falling or balls rolling from hills. It is purely an effect of relativity. The question, then, is the following: If we start with the universe almost exactly at the equilibrium point of the potential (per my analogy, the pencil vertical), how many separate horizon volumes are created by the doubling in size before inflation ends (the pencil falls)?

The answer relies on a competition between two rates: how fast the universe is doubling in size versus how quickly the unstable state decays, analogous to how long it takes the pencil to fall from the vertical to the horizontal. The faster the rate of doubling or slower the fall to equilibrium, the more horizons—separate universes—result. Remarkably, in inflation, these two timescales are set by one number: the expansion rate. The faster the expansion, the shorter the doubling time. But the friction that creates the slowly rolling field *also* comes from expansion: the faster the expansion, the slower the rolling of the field and decay to a stable equilibrium. Both of these effects work in the same direction: more doubling. If we only consider the classical behavior of the field, the duration of inflation must always be finite and therefore the number of separate universes produced must likewise be finite, although it can be very large. This changes when we consider the effect of quantum mechanics.

I argued above that quantum fluctuations—for example, in a pencil balanced on its tip—act to push a system away from an unstable equilibrium. When we include the size doubling that happens during inflation, however, something new happens. This is because quantum fluctuations—inevitable due to quantum uncertainty—are as likely to push the system toward equilibrium (i.e., up the hill) as they are to push it away from equilibrium (figure 5.3). Like the doubling rate, the size of quantum fluctuations depends on the expansion rate: the higher the expansion rate, the larger the quantum fluctuations in the inflaton field. This is referred to as *quantum diffusion*.

There are then two sets of dynamics at play. The first is *classical* evolution of the field always down the hill from higher to lower potential. The second is quantum diffusion, which is random; the field is as likely to jump back up the hill as it is down. Near the end of inflation, classical evolution dominates and quantum effects are small. This is exactly the period during which quantum fluctuations create the primordial perturbations we see within our horizon today. Yet earlier in inflation, the field rolls so slowly and the expansion rate is so high that quantum diffusion dominates over classical evolution. Linde first showed in 1986 that when this is the case, the doubling in size of the universe continues *indefinitely*, resulting in an infinite number of causally disconnected horizons—a process that Linde referred to as "infinite self-reproduction."[4] Because of its quantum nature, infinite self-reproduction is a statistical process. The field will relax to its stable equilibrium, ending inflation and creating a universe like our own, an infinite number of times. But for each universe in which inflation ends, there will be an infinite number of new inflating horizons generated via the self-reproduction caused by the size doubling of the inflating space. This process continues forever, without bound. An analogy to the eternally inflating universe is bubbles in a glass of beer (figure 7.1): each bubble in the beer is a universe like our own, with the edge of the bubble universe expanding outward at the speed of light. In between bubbles is vacuum-dominated, inflating space, expanding exponentially and pulling the bubble universes apart from one another *faster* than the speed of light.

Remarkably, each bubble universe like our own appears finite to an outside observer, but appears *infinite* to an observer (like us) inside the bubble.[5] We see the "wall" of the bubble not as a boundary in space but rather as a boundary in time—the moment in our past when eternal inflation ended and classical evolution commenced (figure 7.2). This takes a certain amount of imagination to visualize. Note in particular that the "bubbles" in figure 7.1 are not the horizons depicted in figure 3.2; the observable universe represented by our cosmic horizon is a tiny patch of the infinitely large universe contained within a *single* bubble in figure 7.1. (Despite its quantum mechanical origin, the multiverse generated by eternal inflation is in no way related to the

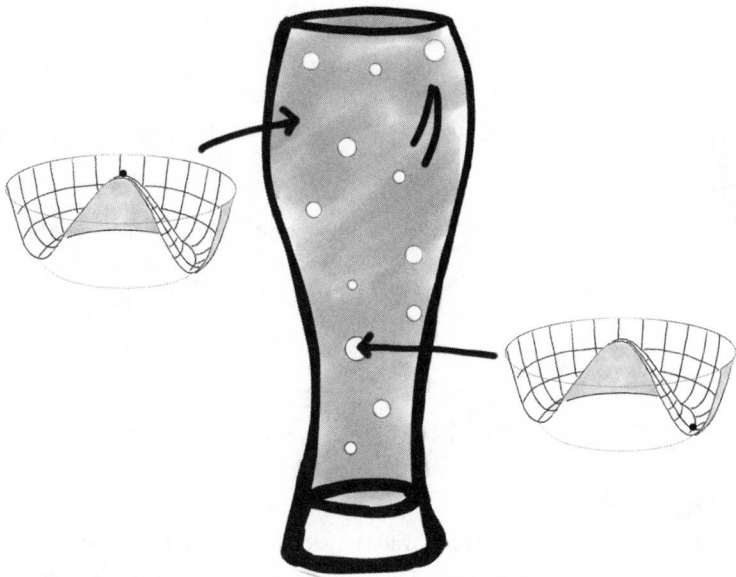

Figure 7.1
Bubble universes in an inflating space, like bubbles in a glass of beer. Each bubble of stable vacuum (*white circles*) expands outward at the speed of light, but the inflating space between them, still in the unstable state, pushes them away from each other faster than light.

"many worlds" interpretation of quantum mechanics proposed by physicist Hugh Everett.[6] The bubble universes here are physically real.)

Including quantum mechanics in our picture of inflation therefore has two important physical consequences: the generation of primordial density perturbations, and eternal inflation. These are in fact intimately related. Eternal inflation occurs when the average size of quantum fluctuations in the inflaton is larger than the amount of classical field evolution during one doubling

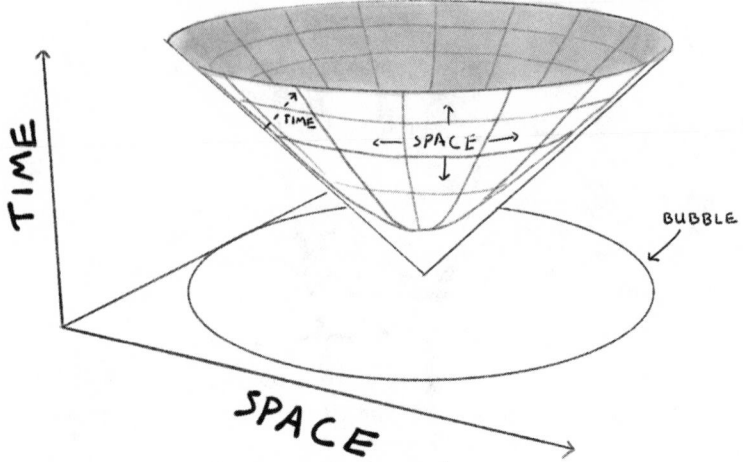

Figure 7.2
Embedding an infinite space inside a spatially finite bubble. The bubble, expanding at the speed of light, traces out a light cone in spacetime. We create infinite spatial surfaces (*shaded grid*) inside a spherical bubble by nesting a hyperbola inside the light cone, effectively borrowing time and redefining it as space. Time points to the inside of the cone, and the bubble wall lies in the past from the perspective of an observer inside.

time of the universe. At the same time, the *ratio* of the amount of quantum and classical evolution of the field is the number that determines the size of the primordial density perturbations (chapter 5).[7] When the ratio is less than one, classical evolution dominates, and when the ratio is greater than one, quantum evolution dominates. The 10^{-5} amplitude of primordial perturbations in the CMB tells us that classical evolution dominated during the period late in inflation when these perturbations were generated. Knowing what happened earlier in inflation requires extrapolation beyond what we can observe, which depends on our construction of models for inflation in particle physics. We cannot *know* any one of these models is correct; we can only say which are consistent with the data, and which are not. With this huge caveat in mind, we can say that *most*, if not all, reasonable models for inflation in particle physics inevitably include eternal inflation. This includes models that incorporate instability due to symmetry breaking, akin to the Higgs mechanism in the standard model of particle physics as well as models based on modifications to general relativity, such as Starobinsky inflation and the related models based on inflationary "attractors" in supersymmetry, or indeed any model consistent with observation in which the inflaton evolves away from an unstable equilibrium point.[8] Eternal inflation is the rule and not the exception.

This is a curious state of affairs; in constructing a picture of the early universe that explains its current observed properties, we find that almost any model results in the prediction that inflation runs out of control forever into the future and there should be an *infinite* number of universes like our own, embedded in a larger, eternally self-reproducing inflationary space-time. The old picture of a single big bang universe arising out of an initial singularity is all but obliterated, replaced by an almost incomprehensibly larger cosmos, reminiscent of the multitude of universes referenced in Hindu cosmology: "There are innumerable universes besides this one, and although they are unlimitedly large, they move about like atoms in You."[9] More disturbing still, the inflationary multiverse lies outside our cosmic horizon, forever out of reach of any conceivable observation or experiment. If this is indeed the case, the multiverse lies in a gray zone at the boundary of—or perhaps a little beyond—what can be called science at all.

How severe a problem this represents depends on the larger theory into which we embed inflation. In the simplest scenario in which the vacuum state of the universe—the minimum of the potential—is unique, the existence of an inflationary multiverse represents few conceptual difficulties. Just as our cosmic horizon represents a finite patch of a spatially infinite universe, in the multiverse picture, that infinite universe is one of an infinity of other identical universes. In both cases, there are unimaginably vast reaches of space and time that we are *in principle* forbidden from observing. Nonetheless, this infinity of infinities preserves the Copernican principle since the multiverse, like the universe beyond our cosmic horizon, is on average the same everywhere, and we hold no privileged position.

String theory changes this picture dramatically.

The Multiverse in String Theory

When seeking to understand the physics of the very early universe, the ultimate goal is to move beyond a qualitatively descriptive or *phenomenological* picture of inflation, and incorporate it into a more fundamental theory, likely involving the unification of quantum mechanics and gravity. We do not have such a theory in hand, but one of the most widely studied candidates is string theory, and it is illuminating to consider the specific example of string-based inflation. This is an uncertain undertaking because the properties of string theory are themselves uncertain, and there is currently a robust debate in the fundamental physics community as to whether or not string theory even admits inflation as a self-consistent solution.[10] Despite this, there is a rich literature of inflation models constructed in string theory, and I will proceed here from the viewpoint that such models are in fact consistent with a complete theory of quantum gravity.[11] I will return briefly to the alternate viewpoint at the end of this chapter and when I discuss alternatives to inflation in chapter 8.

If string theory is in fact the correct framework for the unification of gravity and quantum mechanics, a so-called theory of everything, then it is reasonable to expect that the physics responsible for cosmological inflation will be contained somewhere within the theory. Indeed, string theory provides not just one but rather *many* candidates for the inflaton field in the form of fields known as *string moduli*. To understand modulus fields, we need to understand a related concept, compactification, first proposed by Oskar Klein in 1926 to explain why electric charge comes in discrete units.[12]

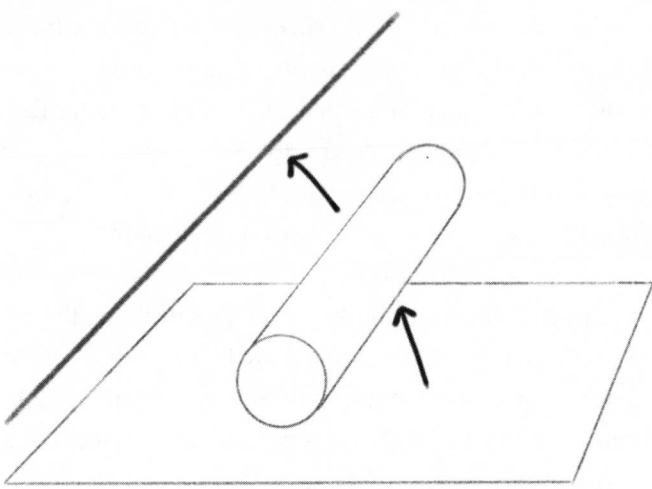

Figure 7.3
A schematic diagram of compactification, in which one dimension is rolled up into a compact form. If the compactified dimension is made small enough, the surface appears lower-dimensional to an observer.

Compactification arises in any theory in which the number of dimensions of space and time is larger than the four (three space plus one time) that we observe in nature. If a theory contains extra dimensions, those dimensions must be hidden in some way or else the theory cannot match observed reality. One way to hide an extra dimension is to roll it up into a tube with a circumference that is microscopically small (figure 7.3) so that the extra dimension becomes invisible to macroscopic experiments. This "rolling up" of extra dimensions is called *compactification*. In the case of string theory, there are a *lot* of extra dimensions to hide, or six in the case of (ten-dimensional) superstring theory. Compactifying one extra dimension is simple, but compactifying many is extremely complex, with many possible configurations. In string theory, the geometry of the compactification maps uniquely to a choice of symmetry for the resulting theory of particles and fields at low energy. The standard model, with strong, electromagnetic, and weak forces, corresponds (in principle) to one choice of compactification of the extra dimensions of string theory. (This is a conjecture; no such explicit construction of the standard model has been directly demonstrated.) The number of possible compactifications—and therefore the number of possible resulting "laws" of physics at low energy—is huge, estimated to be somewhere around 10^{500}.[13] This is a gigantic number. For comparison, the number of possible legal positions in chess is estimated to be around 10^{50}, the number of atoms in the universe is about 10^{80}, and the number of legal positions on a standard nineteen-by-nineteen board in the game of Go is around 10^{170}.

In a stringy picture of the very early universe, the cosmos begins in a "quantum gravitational" state, presumably of higher dimension, and in no way even superficially resembling our familiar universe of three dimensions of space and one of time. During this phase, space and time themselves would have been quantum mechanical in nature, with properties described by quantum uncertainty rather than classical notions of causality. Little or nothing is understood about how such a cosmic state would be constructed, or how it would behave. Without such a description, we are left to speculate that—somehow—this higher-dimensional quantum structure relaxed into one of the 10^{500} possible compactifications (also referred to as *vacua*) of string theory, resulting in *our* universe of three plus one dimensions, monotonic evolution in time, and classical causality. This process of compactifying from a high- to the present low-dimensional state is a dynamical one, described by the evolution of a large number of quantum fields from unstable states to stable equilibriums. One way to think of these fields, called *moduli* (singular *modulus*), is simply as a set of parameters or a collection of numbers that together depict the particular way in which the high-dimension space is folded up into a low-dimension space. One (or perhaps more) of the string moduli can serve as the inflaton; as the decay of the universe from an unstable high-dimensional state into a stable low-dimensional vacuum is exactly the sort of instability that can drive the exponential expansion of inflation. Many specific constructions of such symmetry breaking in string theory have been proposed.[14] An analogy to visualize this is based on the simple picture I introduced earlier of the inflaton field as a ball rolling off the top of a hill (an unstable state, like a pencil balanced on its tip) and eventually coming to rest in a valley (a stable state, like a pencil lying on its side). In the string picture, there are many modulus fields—the balls rolling on hills—and a "landscape" of many peaks and valleys in a space with many dimensions, with each valley representing a stable compactified state. This collection of dynamical moduli and stable compactifications is called the *string landscape*, a term credited to physicists Leonard Susskind and Lee Smolin, who used it as an analogy to the "fitness landscape" of evolutionary biology in his 1997 book *The Life of the Cosmos*.[15]

Out of such an enormous number of possible stable compactifications of the extra dimensions of string theory, how does the universe know to select *our* universe, with three plus one dimensions, and the symmetries and particles of the standard model of particle physics? The answer is that it doesn't. Any one of the 10^{500} possible string vacua are equally likely, and selecting the one that corresponds to the universe we observe is vanishingly *unlikely*. It is here that the inflationary multiverse provides a rescue of sorts. If inflation leads to the production of an infinite number of causally separate universes, then inevitably *all* the possible vacua of the string landscape will be populated an infinite number of times. We live in the one we do because we *can* live in it. This brings us, inevitably, to the anthropic principle.

The Anthropic Principle

The eternally inflating universe is something like a glass of beer, or a smooth fluid full of tiny bubbles. Inside each bubble is an infinite universe. The fluid itself is infinitely large and contains an infinite number of bubbles—an infinity of infinities. Everything we see or ever can see is inside one bubble, but there exists an infinity of other universes, each forever separate from ours. (Sometimes two of these bubbles can collide, and astronomers are now looking for evidence of such collisions, so far without success.)[16] Even stranger, as we have seen, string theory suggests that the laws of physics will be different in each bubble. Different "laws" of physics can come about because of variations of fundamental numbers in physics. The standard model of particle physics contains nineteen "free" parameters, which are not explained by the theory and must be determined by experiment. Extensions to the standard model based on supersymmetry contain hundreds. For example, the *coupling constant* of a particle is a dimensionless number that expresses the strength of a fundamental force in nature. The coupling constant associated with electromagnetic interactions is called the *fine structure constant* and has a value of 1/137. Particles such as the Higgs boson interact not only with other particles but with themselves as well, and this "self-coupling" determines the strength of the interaction. The self-coupling of the Higgs has not yet been well measured, and it will likely require a successor to the LHC to do so.[17] Remarkably, the CMB allows us to measure the self-coupling of the inflaton field; for inflation taking place near 10^{15} GeV, the corresponding self-coupling of the inflation field is around 10^{-14}—a tiny number compared to the electromagnetic coupling constant. Why is this number so small? There are many other examples of such "fine-tuning" in nature. An especially famous one is the value of the cosmological constant expressed in Planck units, $(\Lambda/M_P)^4 \sim 10^{-120}$, which is surely the weirdest small number in the history of physics. Nobody has the faintest idea why it takes the value it does.

The so-called anthropic principle suggests that the values of fundamental constants are what they are because of a selection effect: if they were different, we would not be here to measure them. In its weakest form, the anthropic principle is a tautology: our existence necessitates physical laws that allow it. In stronger form, the idea is hugely controversial. Two developments in theoretical physics have led to a recent interest in the anthropic principle in physics:

1. The concept of a multiverse—that what we call our universe is actually one of many, which is a natural consequence of cosmological inflation.

2. The string landscape, which is the realization that string theory contains an enormous number of stable states for the vacuum, estimated to be as large as 10^{500}. Each different stable vacuum in the landscape will have different symmetries and couplings for fundamental particles—for all intents and purposes, different laws of physics.

If these two ideas are correct, then the consequence is that there is a huge population of universes out there, each with its own "fundamental" physical laws. In that plethora of options, however, life will find itself existing only in universes with physical constants compatible with the existence of life. This will be true even if the probability of finding a universe compatible with life is vanishingly small. Hence we do not need to explain why the fine structure constant, for example, is what it is. In fact, there is no explanation. I take issue with the anthropic argument for three reasons:

1. It is not a scientific argument because it is neither predictive nor testable. We cannot, even in principle, observe the distribution of fundamental constants in other universes, or even determine whether or not such universes exist.

2. The anthropic principle contains no prescription for deciding which properties of the universe are anthropically selected, and which are not. Which constants do we allow to vary from universe to universe, and which are fundamental and do not vary?

3. The anthropic principle is based on a very narrow concept of "life" and what physical conditions make life possible. We currently know of only one example of an environment containing life, which introduces a massive bias in how we conceive of what makes life possible.

Let us discuss each of these objections in turn.

First, the anthropic principle is not science in the sense that science is the process of finding natural explanations for things we do not understand. By contrast, the anthropic principle is an assertion that no explanation of certain phenomena is possible, even in principle. Consider the fine structure constant again as an illustration. A devotee of religious creationism might hold that the apparent fine-tuning of α is evidence for deliberate design in the universe; it is as if the fundamental laws of physics were deliberately set to make our existence possible. The anthropic principle likewise links the value of the fine structure constant to our existence; because carbon-based life would not exist in a universe with a different value for α, the apparent fine-tuning is evidence for anthropic selection in a multiverse. It is sometimes argued (for example, by Nobel Laureate Steven Weinberg) that the anthropic principle is in fact just Charles Darwin's idea of natural selection writ large; just as life itself arises through random mutation, the laws of physics themselves are manifestations of a random process.[18] It is true that natural selection, like the anthropic principle, contains randomness as a central feature. Natural selection, though, contains a crucial ingredient that the anthropic principle lacks: a measure of fitness. Successful traits reproduce more efficiently than unsuccessful ones. The efficiency of replication is a measure (in fact, the sole one) of fitness in evolution. By contrast, the anthropic principle proposes no such measure of the "fitness" of the universe. Quite the opposite, it proposes on fundamental grounds that there is no measure at all and grafts on a selection effect a posteriori. (Attempts have been made at constructing a pseudo-Darwinian measure of fitness for cosmology, most notably by Smolin, but the results are less than compelling.)[19] The anthropic universe is ultimately wasteful and undirected, and we occupy a vanishingly unlikely oasis in an inconceivably huge desert of mute, lifeless universes. Perhaps this is true, but when all we can see is that single oasis, it is an incredible stretch to conclude that the desert must exist.

The second objection to the anthropic principle lies in the assumption of what is allowed to vary from universe to universe, and what is not. For instance, why do the fine structure or cosmological constants randomly vary from universe to universe, but π or Napier's constant do not? In a casual straw poll of my colleagues in cosmology, I have asked the question, If you can imagine a universe in which the laws of physics are different, can you imagine a universe in which the laws of mathematics are different? The answer I receive overwhelmingly is "no." Mathematics is believed to be universal in a way that physical law is not. We can accept the idea of a universe in which the strength of electromagnetism is different, but we cannot even imagine a universe in which there are no prime numbers. The problem is that nobody really knows where to draw the line. Which properties of our universe are truly "universal" and true in every instance of the multiverse, and which are variable and subject to anthropic selection? What principle guides the distinction? This is no small problem. Consider the odd fact that

$$9876543210/1234567890 = 8.0000000073.$$

Why is this odd fraction equal to an integer to eight significant digits? How can we explain such a strange "fine-tuning" in the universe? Of course there is a mundane mathematical explanation for this apparent coincidence, which I leave for the reader to discover. Similarly, other things that might appear to be odd coincidences or fine-tuning may in fact have perfectly understandable explanations. The fact that we do not have a perfectly understandable explanation for a particular "cosmic coincidence" cannot be taken as evidence that no such explanation exists. Anthropic explanations for such coincidences amount to just so stories to explain away our ignorance.

The third objection to the anthropic principle concerns the underlying assumption about the conditions necessary for life. Ironically, although the anthropic principle relies on universes being inconceivably numerous, it simultaneously depends on life being exceptionally rare. If life were common, there would be no selection effect that would bias our observations toward universes with certain well-defined physical properties. The problem is that we have no idea how common life is in our own universe, much less how common it is likely to be in a multiverse with a wide distribution of physical laws. We have one example of one planet with life, and the logic of the anthropic principle requires the extrapolation that any life must be more or less similar to us. We are biochemical machines, built out of atoms, with a structure determined primarily by electromagnetism. In a cosmic sense, we are incredibly fragile; we can only survive in a very special environment, with a very narrow range of temperature, protection from cosmic radiation, liquid water, and so forth. Life like us, even if it is widespread in the universe, will only be found on little rocks that happen to be just the right distance from stable, long-lived stars. In a universe with no stars, no little rocks, or no complex molecules, there will be no life like us. We are extraordinarily fine-tuned.

But is "life like us" the only possibility for how one might build life? There is no particular reason to think so. Natural selection does not contain biochemistry as a fundamental assumption. Any system with sufficient complexity to self-replicate is sufficient. Furthermore, natural selection tells us that life will optimize itself to suit the environment in which it evolves. We are highly specialized to survive on a little ball of rock with liquid water because we *evolved* on a little ball of rock with liquid water. Life that evolves in radically different circumstances will be specialized to survive in those circumstances and might not look anything even remotely like us. We simply have no data on the environments in which other life-forms can (or do) exist. One thing that is being realized in modern biology is that even our narrow, carbon-based definition of life is astoundingly flexible; so-called extremophile life exists in environments so harsh that they were previously assumed to be lifeless. The one lesson that we are learning over and over is that life is more adaptable as well as ubiquitous than we ever previously expected. Bacteria may well be widespread in the solar system, not just on Earth. My guess (and it is only a guess) is that someday, when we understand life better, we will find that life in the universe is not rare but rather is ubiquitous and fills a huge variety of wildly implausible niches. Maybe we just haven't looked for it in the correct way. Perhaps there are organisms composed only of dark matter or black holes, based on gravity instead of electromagnetism. Why not? (Physicist Lisa Randall, for example, has proposed the possibility of complex "shadow" life existing in a "dark sector" that does interact directly with normal matter.)[20] Such life-forms could populate a universe that a naive application of the anthropic principle would rule out as "suitable for life." The *known* unknowns are many; the *unknown* unknowns are limitless.[21] If we embrace the idea of the multiverse, any environment with sufficient complexity to develop life will be replicated exponentially many times, meaning that if it is possible for natural selection to take hold in a given environment, it inevitably will. The idea of a multiverse is by its nature incompatible with the idea of the rarity of life, but the anthropic principle requires both in order to make any sense.

Even the question of what we mean by "likely" or "unlikely" is more complicated than it might at first seem. Precisely *how* unlikely is our universe? In the case of an infinitely self-reproducing cosmos like that of the inflationary multiverse, this question leads to a deep mathematical ambiguity known as the *measure problem*. Consider a simple analogy for the selection of a stable vacuum in the string landscape. Imagine we have a box that contains a hundred marbles, fifty of which are black and fifty of which are white; the different colors will represent different string vacua or choices of low-energy physics. If you reach into the box and randomly select a marble, the probability that you will select a black marble is 50 percent, and likewise the probability that you will select a white marble is 50 percent. As long as the number of marbles in the box is finite, the probability of selecting one color or another is well defined. Now imagine an *infinite* collection of black and white marbles, arranged in alternating colors (black/white/ black/white . . .) extending forever in both directions (figure 7.4). What is the probability of picking one color or the other in a random selection? It's tempting to say, just as in the case of fifty black and fifty white marbles, that the probability of selecting either color is 50 percent, but this is not so. To see why, imagine rearranging the marbles by swapping one for another in an infinite series of swaps, until the balls are arranged with two black marbles for every white one (figure 7.4). Since the numbers of both black and white marbles are infinite, it

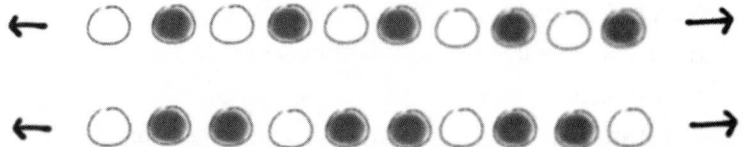

Figure 7.4

An illustration of the measure problem. Imagine the two sets of black and white marbles above extending infinitely in both directions. The two sets are then identical since one can be transformed into the other by rearrangement. Because of this ambiguity, the probability of selecting a black or white marble at random cannot be defined.

is perfectly consistent to reorder the series in this way! Now what is the probability of selecting a white marble? The answer would seem to be one-third. But the sets are identical! In fact, we could rearrange the marbles to give any apparent probability we wish, thus telling us that the probability is actually undefined; there is no unambiguous way to assign a selection probability (called a *measure*) on the infinite sets of black and white marbles. The inflationary multiverse suffers from exactly the same problem of a lack of a consistent measure on the space of outcomes; here the balls in the box are analogous to bubble universes, and the colors of the balls are analogous to choices of vacuum state in the string landscape. It is not just hard to calculate the probability of a universe like ours being produced out of the many possibilities in the string landscape but instead impossible to even *define* what we mean by probability in such a case because the multiverse created by eternal inflation is unbounded, with every possible outcome happening an infinite number of times. In this sense, the entire question of the likelihood of our universe is meaningless.

Ultimately, the anthropic principle depends in an integral way on the assumption that we are in some deep and fundamental way special, or what you might call "carbon exceptionalism." Only a vanishingly rare set of circumstances will create beings such as us so that we must have a universe finely tailored to our existence. Not only is this an embarrassingly anthropocentric notion, it clashes with the foundational idea dating in its earliest form to Copernicus: the cosmological principle. Copernicus's radical idea was that the Earth enjoys no special position in the cosmos and is just one of many planets. Modern cosmology extends this by realizing that the sun is but one of many stars in a galaxy that is one of many galaxies in a universe that may be just one of many universes. The cosmological principle, in a nutshell, is that we are ordinary, and it is this very ordinariness that forms the fundamental organizing idea of modern cosmology. This tension was fully realized by the physicist Brandon Carter, who originally proposed the anthropic principle in a 1974 essay:

> [The anthropic principle] consists basically of a reaction against exaggerated subservience to the Copernican Principle. Copernicus taught us a very sound lesson that we must not gratuitously assume that we occupy a privileged central position in the universe. Unfortunately, there has been a strong (not

always subconscious) tendency to extend this to a most questionable dogma to the effect that our situation cannot be privileged in any sense.[22]

Perhaps this is so. But we likely cannot *know*, even in principle. The inflationary multiverse places us at—or perhaps beyond—the boundary of what we can legitimately think of as scientific hypothesis, and this is a bitter pill for any scientist to swallow.

New ideas in string theory are challenging the concept of the string landscape altogether. Instead of the "top-down" approach of beginning with a theory of quantum gravity and attempting to compute its consequences, the alternative is to build "bottom up," starting with known physics at low energy and attempting to constrain the properties of any possible theory at high energy. This leads to the idea of the string "swampland," which is the set of theories of particles and fields defined at low energy that are conjectured to be incompatible with any consistent theory of quantum gravity.[23] From this viewpoint, the 10^{500} branches of the string landscape are almost all dead ends. The anthropic argument is then reduced to a phantom, an irrelevant ghost haunting the vast swampland of unrealizable string vacua. Most important for the discussion here, recently proposed conjectures postulate that any vacuum energy long lived enough to support cosmological inflation beyond a few doubling times must *also* lie in the swampland.[24] If this is true, it means that the infinite multiverse of eternal inflation is almost certainly an artifact of an inherently ill-posed theory, as are all models for inflation involving a single field.[25] Note, however, the word *conjecture*; swampland proposals—at present—rely entirely on a set of consistency conjectures that are theoretically well motivated, but are supported by neither rigorous mathematical proof nor empirical evidence. Time will tell, but it is reasonable to be skeptical of conclusions relying solely on the blunt, and blind, instrument of conjecture. Nature is entirely indifferent to how we think it ought to work.

8
Just So Stories

At last things grew so exciting that his dear families went off one by one in a hurry to the banks of the great grey-green, greasy Limpopo River, all set about with fever-trees, to borrow new noses from the Crocodile. When they came back nobody spanked anybody any more; and ever since that day, O Best Beloved, all the Elephants you will ever see, besides all those that you won't, have trunks precisely like the trunk of the 'satiable Elephant's Child.

—Rudyard Kipling, *Just So Stories*

I have argued that inflation is a specifically predictive and remarkably successful scientific theory. Inflation provides a compelling explanation for the observed geometric flatness and overall homogeneity of the universe as well as the presence of a nearly scale-invariant, Gaussian spectrum of superhorizon perturbations. It is tiny initial fluctuations in the density of the universe that are—via gravity—the origin of all cosmic structure. Inflation also makes specific predictions that have yet to be confirmed, such as the presence of a background of primordial gravitational waves, so-called consistency conditions on the form of the gravitational wave spectrum, and on the predicted small deviations from perfectly Gaussian statistics. The cold, empty, exponentially expanding space-time of inflation *precedes* the onset of the hot big bang universe and in this sense provides an explanation for the origin of the LCDM cosmology. LCDM, although almost certainly incomplete, is a "concordance" model that fits multiple independent observations, supplying a precise framework for future precision tests of cosmology.

But is inflation *right*? We saw in chapter 7 that if we calculate the consequences of the inflationary picture of the early universe, we arrive at the bizarre conclusion that inflation is most likely future eternal, forever spawning an infinity of new universes, each forever cut off from contact with its siblings by inescapable limits of causality. This strange—and fundamentally untestable—picture calls into question inflation's status as a fully scientific theory, and recent theoretical conjectures argue that inflation resides in the swampland of theories incompatible with a well-formulated theory of quantum gravity. In addition, despite its broad explanatory power, inflation *fails* to explain key features of the universe, and itself contains mysterious apparent inconsistencies, which I will discuss in the next section. This has left some cosmologists dissatisfied, leading them to search for alternatives. In this chapter, I will survey the larger picture of inflation in the context of alternative theories, leading to inescapable and perhaps unanswerable questions about the limits of scientific knowledge.

What Inflation Doesn't Explain

We have seen that inflation is generically (but not inescapably) *future* eternal. That means that once started, inflation continues forever. The end of inflation in any local patch of space-time, followed by decay of the vacuum energy into particles in the phase of reheating, sets the initial condition for the hot big bang universe. In this sense, inflation happens *before* the big bang, and the end of inflation replaces the initial singularity that was previously understood as an inevitable feature of the big bang. But does inflation replace the initial singularity or does it *displace* it? Put another way, if inflation is future eternal, can it also be in some sense *past* eternal?

This is a subtle question because of the malleable nature of time in relativity. Time is what we measure it to be, and different clocks measure time differently. None is preferred over any other. Even how we break up space-time into "space" and "time" is (within certain rules) arbitrary; an example of this is the "bubble" universe in figure 7.2, where we borrow from an infinite reservoir of time to turn the interior of a spatially finite bubble into a spatially *infinite* universe. Another illustration, first suggested to me by Nobel Laureate John Mather (during a memorable conversation over beer and pizza at the Buffalo, New York, restaurant Pizza Plant), is this: suppose we use a clock that measures time *logarithmically* rather than linearly. That is, we use a clock that runs extremely quickly in the very early universe and slows down as the universe expands. A logarithmic clock still sees an infinite future as infinite, but it also sees a finite *past* as infinite since the logarithm maps zero on the number line to negative infinity. Relativity tells us that such a clock is just as valid as the standard, linear clocks that we normally use. This presents an apparent paradox: two equally valid clocks, one that measures the initial cosmological singularity to be at a finite time in the past, and another that measures it to be at an infinite time in the past. Which is correct?

To resolve the paradox, we need to find definitions of finite and infinite that do not depend on such arbitrary definitions of time. To do this, we restrict ourselves to a special class of observer: *inertial* observers. An inertial observer is one who is in free fall and experiences no acceleration. Equivalently, an inertial observer is one who is weightless. The path traced through space-time (the *world line*) of such an observer is called a *geodesic* (figure 2.2). The length in space-time along the geodesic path is interpreted in relativity as the *time* measured by that observer, called the *proper time*, and is unambiguously defined. We can then define an infinite space-time as one that extends infinitely for all inertial observers—a property referred to as *geodesic completeness*:

- A space-time is *geodesically complete* if all geodesics in the space-time are both future and past infinite.

The space-time of special relativity diagrammed in figure 2.3 is geodesically complete. By contrast, the big bang space-time depicted in figure 2.4 is geodesically *incomplete*. More specifically, the big bang space-time is future complete and past incomplete, which means that there is at least one observer who measures the initial singularity to be at finite proper time in the past (in fact there are an infinite number).

Now we can ask the question in a more rigorous way: Is the space-time of inflation geodesically complete? If we consider only the set of comoving observers— that is, those who are at rest relative to the cosmic expansion—we find that the exponentially expanding space-time of inflation extends—for those observers— infinitely into the past, similar to the observers with logarithmic clocks in Mather's thought experiment. This may lead one to believe that inflation may enable us to get rid of the initial singularity of the big bang altogether. But for a space-time to be geodesically complete, it must be complete for *all* geodesic paths through the space-time. This is not so. The space-time of inflation was shown to be geodesically past *incomplete* by physicists Arvind Borde, Alan Guth, and Alexander Vilenkin in 2003.[1] There are an infinite number of inertial observers whose past time line is finite, emerging from an initial singularity. This has dramatic consequences; it means that we can push the initial singularity as far back in time as we wish, but we cannot *remove* it. It remains, lurking in the murky past out of which inflation itself arose. We cannot duck the question of what set the initial conditions for inflation itself.

Those initial conditions must be specific. I argued (figure 4.2) that inflation provides a solution to the horizon problem—that is, an explanation of why the universe is homogeneous on scales far larger than the size of the cosmic horizon in early times. The idea is that a small, homogeneous patch smaller than the horizon can be stretched by the exponential expansion of inflation to superhorizon size, explaining this otherwise unexplained large-scale homogeneity. This seems to be a reasonable assumption, but it doesn't completely work. Physicists Tanmay Vachaspati and Mark Trodden showed in 1998 that inflation cannot occur unless space is *already* homogeneous on scales larger than the horizon size.[2] This presents a definite problem for inflation, but it is not completely clear how severe it is. Vachaspati and Trodden's proof demonstrates that the scale of homogeneity must be larger than the horizon, but it does not say how *much* larger, and it definitely does not need to be exponentially so. As University of Chicago physicist Michael Turner puts it, inflation may not solve the horizon problem, but it does make it twenty orders of magnitude better! It is conceivable that such homogeneity could arise at random in some sort of chaotic, preinflationary state, and it need only happen once.[3] Recent progress testing this has been made using numerical relativity, but it is an extremely difficult question to address in general.[4] While we have a good understanding of inflation itself, we have *no* theory describing physics prior to inflation. We do not even know the most basic properties of this primordial universe, such as its dimensionality, and it is likely that any understanding of the preinflationary universe will require a quantum theory of gravity, which remains stubbornly elusive. The preinflationary universe remains terra incognita.

While the problem of initial conditions may seem to undermine the basic rationale for inflation—we have merely taken the initial singularity of the big bang and pushed it back to an earlier, even more poorly understood epoch—it remains a purely theoretical issue. This is because any trace of those initial conditions is erased by the unbounded exponential expansion that takes place during inflation itself, leaving (for all practical purposes) no observational trace behind. Because inflation generates perturbations via expansion, shorter wavelength quantum modes are redshifted out of the horizon later in inflation, and longer wavelength modes exit the horizon earlier on. This means that when we look, for example, at the CMB, we are seeing modes that were stretched to superhorizon scale very near the *end* of inflation. Quantum modes that exited the horizon earlier remain larger than our horizon size today and are unobservable. Any traces of physics near the beginning of inflation are *vastly* larger than our observable universe. This is especially true if inflation is eternal since in that case, there is no lower bound of how much inflation might have occurred between its initial state and the time when the vacuum decayed in our local bubble of the multiverse, ending inflation. The span of time between the two could be unimaginably immense. This property of inflation as a powerful "cosmic eraser" of initial conditions means that as long as inflation got started *somehow*, at the end of the process the result is a universe that looks a lot like ours. This property of inflation—that it is an *attractor*—is a key feature of its success as a description of nature. The late universe is almost completely insensitive to the details of the early universe.

The cosmic "blank slate" created by inflation is a central feature of the theory, but it comes attached with a new problem, which arises from an inconsistency between the sheer amount of expansion required for inflation and the scale of quantum gravity, called the *Planck length*. We can estimate the Planck length in exactly the same way we estimated the size of a hydrogen atom in chapter 5, except the relevant force is gravity instead of electromagnetism. Here, via Einstein's famous relation between mass and energy, the uncertainty in the energy of a quantum system can be equated to an uncertainty in the *mass*,

$$\Delta E = (\Delta m)c^2 = \frac{\Delta p^2}{m}.$$

Then we can write the Heisenberg uncertainty relation as

$$\Delta r = \frac{\hbar}{\Delta p} = \frac{\hbar}{(\Delta m)c}.$$

Quantum uncertainty means that there is an energy (and therefore mass) associated with any process that localizes a quantum system to within a radius Δr, and the smaller the radius, the larger the mass. And that mass must source gravity. So we can ask the question, How small must the length scale Δr be for the associated mass Δm to be enough to form a black hole? This is given by the Schwarzschild radius, which relates the radius of a black hole to its mass,

$$\Delta r = \frac{2G(\Delta m)}{c^2}.$$

Combining the Schwarzshild radius with the Heisenberg uncertainty principle results in a minimum length scale,

$$\Delta r = \sqrt{\frac{2G\hbar}{c^3}} \quad 1.6 \times 10^{-35}\,\text{m}.$$

This radius, twenty-five orders of magnitude smaller than a hydrogen atom, is the Planck length; quantum uncertainty tells us that localizing any system in a radius smaller than the Planck length will immediately form a microscopic black hole. This is the length scale at which classical gravity is expected to break down and require a full quantum theory of gravity for a self-consistent description. The fundamental minimum length implied by the Planck scale becomes mysterious indeed when we combine it with the exponentially rapid expansion that takes place during inflation. Where does all that new space generated by inflation come from, if the existence of a fundamental minimum length means that the space can't have existed beforehand?

This question can be made quantitative by considering the generation of primordial perturbations, as discussed in chapter 5. Inflation stretches short wavelength quantum modes into long wavelength modes, exponentially quickly. For every doubling time of cosmic expansion, quantum vacuum modes double in wavelength. We know from the CMB that inflation had to last at least eighty doubling times and almost certainly went on for far longer. Consider that the radius of the observable universe is about 10^{26} meters. The Planck length is about 10^{-35} meters, which means the radius of the observable universe is about 10^{61} times the Planck length, or a factor of about 2^{202}. This means that it only takes a little more than two hundred doublings in scale to stretch a quantum mode with a wavelength equal to the Planck length to the physical size of the *entire* observable universe today! This is more than the eighty doubling times we need to explain the homogeneity of the CMB, but not much more—only about a factor 2½. This means that if inflation went on for even a little more than the absolute minimum required to match the current universe, scales as large as the entire horizon size today would have originated at scales shorter than the Planck length, meaning that a fully self-consistent description of the evolution in quantum modes during inflation would require an understanding of quantum gravity. This is known as the *trans-Planckian problem* and is independent of any of the details of how inflation itself happens. As long as there are more than about two hundred doubling times between the early stages of the inflationary epoch and the universe today, it means that all structure in the current universe started out smaller than the Planck length during inflation and cannot be fully described without a quantum theory of gravity. This means that the process of quantum mode generation that I explored in chapter 5 is in fact not self-consistent and therefore its validity is called into question. There is no universally accepted resolution to this problem, which was first studied in the context of inflation in papers by physicists Jerome Martin and Robert Brandenberger as well as physicist Jens Niemeyer in 2001.[5] The *trans-Planckian censorship conjecture* proposed in 2020 by Alek Bedroya and Cumrun Vafa argues that any theory that permits trans-Planckian quantum modes lies in the swampland of theories inconsistent with quantum gravity.[6]

Is the trans-Planckian problem really a problem? Possibly, and possibly not. A similar version of the problem has been known in the case of Hawking radiation from black holes since the mid-1990s in work by physicist William Unruh.[7] The history of physics is rich with calculations that appeared to be internally inconsistent, yet turned out to be entirely valid when a deeper understanding became available—a famous example being the "ultraviolet catastrophe" in the classical electromagnetic description of the spectrum of a black body. We know that quantum field theory—the theory we use to calculate the perturbations in inflation—is incomplete since it is incapable of explaining the absence (or presence) of vacuum energy from zero-point quantum modes in the universe. Yet quantum field theory yields exquisitely precise predictions for quantum processes, such as the "g factor" associated with electron spin, which is known to an accuracy better than one part per trillion. What we can say is that once a quantum mode being stretched by inflationary expansion has a wavelength much *larger* than the Planck length, its evolution will be governed by quantum field theory. Since the horizon size during inflation is typically at least ten thousand times the Planck length, standard quantum field theory provides a consistent explanation for the behavior of vacuum fluctuations while they are larger than the Planck length, but still much smaller than the horizon during inflation. Therefore we can map our uncertainty about quantum gravity into a *boundary condition*, or the initial state for the quantum mode, and it is reasonable to expect that the physical picture presented in chapter 5 is valid in at least an approximate sense.[8] One can only make a rough estimate of the expected deviations from scale invariance caused by quantum gravitational effects, but probably a good guess is that the difference should be of the same order as the ratio between the Planck length and horizon size, or well below the percent level.[9] Searches for predicted signatures of trans-Planckian effects in the CMB have so far been negative.

I have identified three issues that potentially represent foundational problems for inflation as a valid model for the early universe:

- Geodesic incompleteness
- No theory of initial conditions
- Trans-Planckian perturbations

(It is often argued that a "lack of predictivity" should be included in this list, but I have argued in chapter 6 that this is not in any reasonable sense a problem for inflation.) What are we to make of this list? All these uncertainties in the inflationary picture are related to our lack of understanding of how to self-consistently construct a quantum theory of gravity. It may well turn out that once we have a theory of everything ready to hand, these apparent problems will cease to be problems, but we do not know for sure. In the absence of a full theory of quantum gravity, one way to study the self-consistency of inflation is to study alternatives to inflation. What if inflation is the wrong theory? What properties must its replacement have? This is the subject of the next section.

Alternatives to Inflation

Any theory we might wish to propose as an alternative to inflation should explain most, if not all, of the cosmic properties that inflation explains. Geometric flatness and homogeneity could reasonably be taken as boundary conditions, set by some as yet unknown symmetry. Cosmic perturbations are a different matter. Nearly scale invariant, superhorizon perturbations in the early universe are an inescapable observational fact, and any theory that claims to replace inflation must explain the observed primordial perturbations if we are to take it seriously. Scale invariance is ubiquitous in nature, occurring in a wide variety of phenomena, but the superhorizon correlations we see in primordial cosmic perturbations are special because they are difficult to create using causal processes. Inflation solves the problem by the rapid stretching of short wavelength vacuum fluctuations, which grow in wavelength much faster than the cosmic horizon. Is there another way to create scale-invariant superhorizon modes? As long as the universe is in an expanding phase, the answer is a (qualified) "no." In 2011, physicists Ghazal Gesh-nizjani, Azadeh Moradinezhad Dizgah, and I looked at the most general possible conditions for the creation of superhorizon perturbations from vacuum in an expanding universe, assuming general relativity is the correct theory of gravity.[10] We found that generating scale-invariant, superhorizon modes consistent with observation requires one of three conditions to hold at some point in the early universe:

1. Accelerating expansion (inflation)
2. Superluminal wave propagation
3. Cosmic density that (vastly) exceeds the Planck scale

These are the *only* possibilities consistent with general relativity and an expanding universe. The second possibility, superluminal propagation, would require us to abandon the idea of the speed of light as a fundamental speed limit in the universe. The third possibility would mean that whatever explanation one might wish to find for the primordial perturbations would require a quantum theory of gravity. Since we do not have such a theory available, it is impossible to evaluate the consistency of such an option. The only other option is a period of inflation to accomplish the necessary stretching of short wavelength modes to superhorizon scales.

There is, however, an additional loophole. What if the universe was not expanding in the early times but instead was *contracting*? During a contracting phase, the cosmic horizon shrinks versus grows. Quantum vacuum fluctuations also shrink, but it is not difficult to arrange for the horizon size to shrink faster than the overall contraction rate so that quantum modes evolve from subhorizon to superhorizon scales. A 1998 paper by cosmologist David Wands showed that a scale-invariant spectrum of perturbations is generated by a contracting universe filled with dark matter.[11] Yet this "matter bounce" model has two key inconsistencies. The first is that the solution is *unstable*. Any small deviations in the cosmic dynamics tend to be amplified, destroying the homogeneous universe, and instead leading to a phase known as a "chaotic mixmaster."[12] The second inconsistency is that as long as general relativity holds, such models do not undergo a bounce into an expanding phase but rather approach a singularity at late time—a time-reversed version of the big bang singularity in an expanding universe. The instability problem in the matter bounce was addressed in a model first proposed in 2001 by physicists Justin Khoury, Burt Ovrut, Paul Steinhardt, and Neil Turok, and known as *ekpyrotic* cosmology, named in reference to the ancient Greek word ἐκπύρωσις (*ekpurōsis*), referring to a conflagration.[13] As opposed to the rapidly contracting matter bounce, the ekpyrotic universe undergoes a slow contraction, with a more rapidly shrinking cosmic horizon, and avoids most (but not all) of the instability issues present in the matter bounce scenario, particularly chaotic mixmaster behavior.

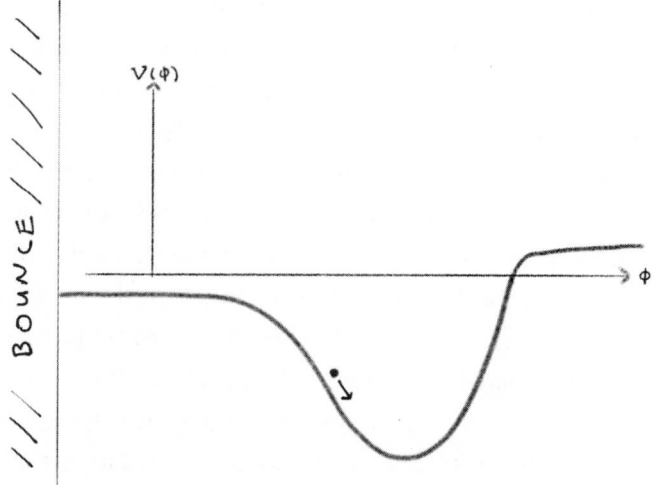

Figure 8.1
The potential surface in ekpyrotic cosmological models.

Like inflation, ekpyrotic cosmology models the contents of the cosmos by a field rolling on a potential surface, except with a Hubble parameter that is negative,
meaning a contracting universe. Unlike inflation, the
potential during the ekpyrotic phase is *negative* (figure
8.1). During inflation, the kinetic energy of the slowly
rolling inflaton field is much smaller than the potential,
leading to exponentially rapid expansion, as discussed
in chapter 4. In ekpyrotic contraction, the kinetic energy
of the field is large, but is almost exactly canceled by the
negative energy associated with the potential so that the
total energy density in the cosmos is almost exactly zero.
Since the Hubble parameter—the rate of contraction—is
proportional to the total energy (kinetic plus potential)

of the field, this leads to an extremely slow phase of contraction, which gradually speeds up as the universe approaches the bounce. Since the cosmic horizon size is proportional to the *inverse* of the Hubble parameter, this means that the horizon is initially large and then shrinks as the universe approaches the bounce, while the wavelength of quantum vacuum modes remains nearly constant. Contrast this with inflation, in which the horizon size is nearly constant, and quantum modes are increasing in wavelength exponentially quickly; in both cases, quantum vacuum fluctuations are initially much smaller than the horizon, but evolve to be much larger than the horizon at late time. Both models solve the "horizon problem" described in chapter 2. Unlike the case of inflation, though, single-field ekpyrotic models do not work. The density perturbations generated by the ekpyrotic field are not scale invariant but instead strongly scale dependent, with an amplitude that becomes larger for shorter wavelength modes, a so-called *blue* spectrum, which is inconsistent with the data. For this reason, ekpyrotic models must postulate perturbations generated by a second field in addition to the field responsible for the ekpyrotic contraction itself.[14] This is a significant shortcoming; ekpyrotic contraction does not automatically produce scale invariance. Rather, it must be engineered into the theory post hoc.

Can ekpyrotic contraction be distinguished from inflation by observation? A crucial distinction between the models is gravitational wave production. In inflation, the amplitude of primordial gravitational waves is proportional to the expansion rate. Since the expansion rate during inflation is large, the production of gravitational waves can be substantial. Similarly in a contracting universe, gravitational wave production depends on the contraction rate. In ekpyrotic cosmology, the contraction rate is extremely small so that likewise gravitational wave production is extremely weak. The result is that any detection of primordial gravitational waves would strongly favor inflation over an ekpyrotic bounce. A *lack* of detection, however, is ambiguous since inflation makes no definite prediction for the amplitude of gravitational waves. They could simply be too weak to be detected. As we saw in chapter 6, the current data place an upper bound on the tensor fraction of about 7 percent.

A main hope for contracting universe models is the possibility of avoiding any sort of initial singularity altogether by making the universe both past and future eternal. A model proposed in 2019 by physicists Anna Ijjas and Paul Steinhardt is of this type. The model is *cyclic*: the universe undergoes slow (ekpyrotic) contraction, followed by a bounce, followed by a period of rapidly accelerating expansion, similar to inflation, before recontracting again and repeating the cycle.[15] An alternative cyclic model, called conformal cyclic cosmology, was proposed by 2020 Nobel Laureate Roger Penrose in

2014.[16] Truly cyclic models for the universe must overcome three fundamental obstacles:

1. Gravitational instability
2. Entropy generation
3. Singularity at the bounce

Gravitational instability is simple: gravity works the same way, whether or not the universe is expanding or contracting. This means that any small perturbations will grow and form structure. Gravity makes the universe *inhomogeneous*. In the case of an expanding universe, this feature of gravity is explanatory; structure formation is responsible for the presence of galaxies, stars, planets, and black holes. Yet in a contracting universe, this same feature is pathological since the presence of collapsed structures like black holes means that a smooth bounce into a homogeneous initial state for the subsequent expanding phase is impossible. Instead, you get a mess. In order to successfully model a homogeneous bounce, it is necessary to explain the absence of structure, despite there being an infinite amount of time before the bounce for structure to form. This is no small problem.

The problem of entropy generation is even more general. Perhaps the most ironclad law in physics is the second law of thermodynamics: disorder or *entropy* in the universe always increases. This has nothing to do with whether the universe is expanding or contracting;

broken glasses do not spontaneously reassemble themselves when the universe contracts! (It is possible to see the problem of gravitational instability as an example of entropy increase; the collapse of matter into a black hole *maximizes* its entropy.) It was recognized as early as 1931 by mathematical physicist Richard Tolman that entropy generation prevents true cyclicity in the universe.[17] Every cycle of a bouncing universe must have a higher entropy than the cycle preceding it. This means that extending cyclic cosmology into the infinite past presents a problem: after a finite number of past cycles, the result is either a state of zero entropy or singularity. Unless one can entirely reverse the arrow of time, including the increase of entropy, it does not appear to be possible to avoid the inevitability of an initial singularity. Ijjas and Steinhardt's 2019 model claims to solve the problem by making the scale of the universe in each cycle exponentially larger than the previous one so that the entropy in any local patch of space remains small. In this case, the entropy from previous cycles does not disappear but instead is simply diluted into ever-larger volumes by a period of accelerating expansion (i.e., inflation) in each cycle.

The third problem to be solved in cyclic cosmology concerns what happens at the moment of the bounce. It is a theorem in general relativity that in the case of a geometrically flat universe, a transition from contraction to expansion must either progress through a singularity or violate a consistency requirement known as the *null energy condition* (NEC). What happens to the perturbations, created during the contracting phase, when the universe passes through the singularity at the bounce? This question cannot be answered as long as the singularity is present and indicates an unavoidable breakdown of the theory. What about the loophole: the violation of the NEC? The NEC states that all particles moving at or less than the speed of light must have a positive (or zero) total energy. Conversely, a violation of the NEC means that there exist particles with *negative* energy. This is not necessarily a problem in classical physics, but in the quantum world in which we live, it is catastrophic. NEC violation means that the quantum ground state of lowest energy does not exist and virtual particles can decay into *real* particles by transitioning to a negative energy state. Since there is no lower bound on the negative energy, this process can continue *infinitely*. This process of decay into infinitely negative energy states is called a *ghost instability* and means that even the vacuum itself is unstable to quantum decay. A universe that violates the NEC instantly tears itself apart via ghost decays. The only way out of the horns of the dilemma—singularity or unstable vacuum—is to abandon general relativity, such as by the introduction of extra dimensions to space-time.

What, then, do we learn by considering the possible alternatives to inflation? First, we learn that they are few, as long as general relativity holds: either inflation, a superluminal ether, or a bouncing universe. The appealing alternative of a bouncing universe on closer examination is revealed to open a Pandora's box of problems more severe than the problems it is intended to avoid: gravitational instability, violation of the second law of thermodynamics, singularities, and quantum instability of the vacuum. This is a steep price to pay to avoid an ultimate beginning to the cosmos! Of course, general relativity could indeed require modification in the extreme conditions of the early universe. But if Einstein's theory holds—we do not know for sure—it would exclude nonsingular classical bounces entirely.

The Beginning

Here, at the end of the story, I return full circle to the first sentence of this book: the universe had a beginning. Can one *really* make such a statement? It is unequivocally true that the universe we inhabit—one made of quarks and electrons and neutrinos and photons, with galaxies and stars and planets and oceans and bacteria and trees and all people we love—did in fact have a beginning, 13.8 billion years ago. We can measure the age of the universe to within 200 million years or so—an astonishingly accurate figure for a process of such scale as the emergence of the cosmos itself into being. So, yes, the universe—*our* universe—really did have a beginning. The observable bubble of our cosmic horizon, which is presumably embedded in a vastly larger structure, is all we can see, and likely all we can *ever* see. This is because the universe today is at the end of a period of transition from dark matter domination to dark energy domination. The acceleration of cosmic expansion commenced when the universe was about half its current age. During the previous period of matter domination and decelerating expansion, our cosmic horizon grew, and more and more of the cosmos beyond our horizon became visible with the progression of time. This is no longer happening. As dark energy comes to dominate and we enter a period of late-time cosmological inflation, we will see less and less of the cosmos (figure 4.1). Exactly the same process of stretching subhorizon scales to superhorizon that happened in the early universe is happening *today*, at a vastly reduced rate. Distant galaxies, which we see today receding from us at less than the speed of light, will eventually be swept out of our cosmic horizon by ever-accelerating expansion. Eventually, if dark energy is sufficiently stable, all the galaxies outside the small, gravitationally bound local group of a few dozen galaxies surrounding the Milky Way and Andromeda will have disappeared from the observable universe, leaving us alone in a dark and empty cosmos.

Our cosmic horizon, very probably, is all we get, and it really did have a beginning.

That beginning, of course, begs an explanation, and modern cosmology's effort to arrive at such an explanation has been the subject of this book. Our effort to understand the beginning has been far more successful than might have been reasonable to expect. This is because we can see and measure the echoes left behind in the tiny imprints in the temperature of the CMB, and less directly, the cosmic web of structure across the vastness of the cosmos. In the inflationary picture, the primordial perturbations in the universe were created at the moment our cosmic horizon itself came into being, in a very real sense *before* the onset of the hot big bang out of which our present cosmos evolved. Inflation, because of this, is a definite and testable scientific theory, and has so far passed multiple nontrivial observational tests of its predictions: flatness, homogeneity, and superhorizon perturbations with a nearly scale-invariant spectrum and almost perfect Gaussian random statistics. There are additional predictions—primordial gravitational waves and consistency conditions—that can and will be more precisely tested in the future. It is by any measure a superbly successful theory.

The picture of the larger cosmos that emerges from inflation, however, is *nothing like* the local universe in which we live, which inflation reduces to a tiny patch of an infinite bubble universe embedded in an infinity of other universes, continuously being pulled apart by the self-reproducing space-time of eternal inflation (chapter 7). In the self-reproducing inflationary multiverse, we are driven inevitably to paradox, such as the breakdown of probability as represented by the measure problem, and by the trans-Planckian problem—the stretching of space from the Planck length to far beyond the cosmic horizon. It is possible that some as yet undis-covered principle forbids eternal self-reproduction, just as relativity elegantly evades causal paradoxes in space-time, but it is just as possible that no such principle exists. It is perhaps not surprising that some theorists have resorted to the desperate measure of the anthropic principle to make sense of this paradoxical cosmos, pos-tulating that life like us is somehow special, and our universe is fine-tuned to allow our existence.

Furthermore, inflation does not dispense with the necessity of a beginning but instead simply pushes the initial singularity far into the murky past of the eternal multiverse, with all traces of its nature infinitely diluted by expansion, lost forever yet still felt, like the grin of the Cheshire cat. There are several possible options for dealing with the problem of the initial singularity in inflation. Singularities in general relativity represent a breakdown of the *classical* theory of gravity, and the the-orem of Borde, Guth, and Vilenkin does not take quan-tum processes into account. It is reasonable to expect that the initial state that led to inflation is quantum gravitational in nature, with classical gravity and infla-tionary expansion arising as an emergent phenomenon. There is at present no consistent theory for such a pro-cess. Other options have been proposed. For example, physicists Anthony Aguirre and Steven Gratton have constructed models in which the initial boundary of inflationary space-time is extended by embedding the inflationary multiverse in a larger universe still, popu-lated by many domains of contraction and expansion.[18]

Such a maximalist solution is of course untestable. The remaining alternative is to abandon inflation altogether and attempt to construct a cyclic cosmology, in which the universe passes through one or more bounces from a contracting phase to an expanding one. I have argued here that such models create more problems than they solve—for example, entropy growth, bounce singularities, and vacuum instability—and are incompatible with unmodified general relativity. Bouncing universes could be ruled out altogether by a detection of primordial gravitational waves.

In the words of Alan Turing, "Science is a differential equation, religion is a boundary condition."[19] The existence of a beginning—a boundary in time—means the end of understanding and is anathema to a cosmologist attempting to scientifically explain the entirety of existence. Yet here we are. This is certainly one of the oldest problems in science, dating at least as far back as Aristotle, who postulated ὃ οὐ κινούμενον κινεῖ, or "that which moves without being moved," as the first cause of existence. Building on this, Aquinas identified Aristotle's prime mover as a sentient, necessary being. The inflationary picture in no sense requires the presence of intention. We may well speculate that the origin of inflation itself lies in quantum uncertainty acting on some primordial, inherently quantum gravitational state in which neither space nor time nor causality have meaning—a universe ex nihilo and uncaused. This is the explanation I favor. But can we *know* that this is true?

Our knowledge is itself limited by the finite extent of our cosmic horizon. Look again at the Planck measurement of the CMB in figure 3.7, particularly the uncertainties on the points, represented by the vertical bars. The points on the left correspond to averages over large patches of the sky (*large angles*), and the points to the right correspond to averages over more numerous small patches of the sky (*small angles*). The key point is that we only have one universe, and one sky to look at, thus meaning that we only have a finite sample of the primordial perturbations. Since the perturbations are a random process, that means that there is an inherent uncertainty from the finite sample size, called *cosmic variance*, which becomes more severe when averaging over larger patches of the sky. This is similar to the margin of error in a political poll; because a poll extrapolates a small number of participants (typically a few hundred or few thousand) to a large population, there is an inherent uncertainty that arises from the finite sample size. This can be seen in the uncertainties shown in figure 3.7: the uncertainties on the large-angle averages—the points on the left—are very large, and become smaller for the small-angle points on the right. This is due to cosmic variance; there are fewer big patches of sky, and more small ones. In the case of a poll, the margin of error can be reduced by interviewing more people, but in cosmology we cannot simply observe more universes. We only have the one. This means that cosmic variance uncertainty is fundamental and cannot be overcome by more precise measurement. It is remarkable that in the Planck data, uncertainties due to the instrument itself are smaller than the cosmic variance uncertainty over the *entire* plot shown in figure 3.7. This means that the Planck measurement of the temperature anisotropy is for all practical purposes *perfect*. Short of traveling to another universe, no more accurate measurement is possible. We can obtain more information by other means, such as by measuring the distributions of galaxies or neutral hydrogen during the cosmic dark ages, but fundamental limits from cosmic variance still apply.

Our knowledge of the epoch of inflation is similarly limited by our finite cosmic horizon. Since shorter wavelength perturbations exited the horizon later in inflation, and longer wavelength perturbations exited earlier, the primordial perturbations we see in the universe today allow us to probe only the very end of the inflationary epoch, corresponding to the final eighty doublings before the end of inflation. Anything that happened earlier than those final eighty doublings is *fundamentally* inaccessible to us. This limit to our knowledge raises the question, Is any theory we construct of the ultimate origin of the universe a just so story, inherently unfalsifiable? Quite possibly. We cannot, even in principle, experimentally probe the initial condition for inflation or the existence of the inflationary multiverse because their relics are forever hidden by our cosmic horizon, just as the horizon of a black hole hides the gravitational singularity within.

How are we to make sense of this?

When Giordano Bruno wrote in *Cause, Principle, and Unity* that "to know the universe is to know nothing of the being or of the substance of the first principle, because it is like knowing the accident of accidents," he was grappling with the same problem, with a remarkably prescient sensibility.[20] Bruno's resolution of the paradox was to adopt a radical expansion of the Copernican principle that the Earth was but one of an infinity of worlds. The Copernican idea provided Bruno with an organizing principle through which to comprehend an unbounded—and ultimately unknowable—cosmos. In inflation, we see Bruno's infinity of worlds extended to an infinity of *universes*. The question of whether the cosmos is a typical outcome of a hypothetical ensemble or tiny region of a *real* one is in this sense irrelevant, and—at least at present—an unanswerable question. The first words of Lao Tzu's ancient book of Taoist philosophy, the *Tao Te Ching*, ring true:

> The Tao that can be told is not the eternal Tao.
> The name that can be named is not the eternal name.
> The nameless is the beginning of heaven and earth.[21]

Our understanding may have fundamentally inescapable limits, and we may ultimately have to learn to live in the presence of the nameless as we seek to comprehend the ultimate origin of the cosmos. There may be no other way.

Acknowledgments

I thank Anthony Aguirre, Daniel Baumann, Katherine Freese, David Kaiser, Eugene Lim, Andrei Linde, Salvatore Rappoccio, Syksy Räsänen, Erik Verlinde, and Mark Wyman for discussions and comments, and Niko Šarčević for being generally awesome. I thank Robert Sperhac for comments on an early draft of the book. I am grateful to my PhD adviser, K. T. Mahanthappa, for getting me started on the path, and the many other mentors I have had along the way. Thank you.

The figures were created by me, unless otherwise noted.

Glossary

Acoustic oscillations: Sound waves in the plasma in the early universe.

Anthropic principle: The principle that physical law is constructed in such a way as to enable the existence of sentient life.

Baryon: In particle physics, a composite particle made up of three quarks. In cosmology, any atom or other particle made up of standard model constituents.

Blue spectrum: A non-scale invariant spectrum with more power at short wavelengths.

B-mode polarization: A helical polarization pattern.

Boundary condition: A set of arbitrarily chosen initial conditions for the evolution of a system.

Causal: The property of cause and effect in relativity by which no signal can propagate faster than light.

Compactification: The "folding up" of extra dimensions of space to make them invisible to physics at low energy.

Cosmic microwave background (CMB): Light left over from the big bang, with a temperature of 2.7 K, first observed by Amo Penzias and Robert Wilson in 1965.

Cosmological constant: A parameter in general relativity describing the energy of empty space.

Cosmological principle: The basic assumption of cosmology that our region of the universe is typical of the universe as a whole. Also sometimes equivalent to the assumption of homogeneity and isotropy.

Coupling constant: A number in a quantum field theory determining the strength of an interaction between two or more particles.

Critical density: The cosmic density corresponding to a flat geometry, equivalent to about 10^{-26} kilograms per cubic meter.

Curvature: The deviation of space-time from flat (Euclidean) geometry.

Dark energy: Material in the universe that does not interact with light or collapse under gravity to form structure. It causes the expansion of the universe to accelerate, and may be a form of vacuum energy or cosmological constant.

Dark matter: Material in the universe that does not interact with light but responds to gravity in the same way as baryonic matter, clumping under gravitational attraction to form structure.

Doppler shift: A change in the wavelength of light due to relative motion between the source and observer.

Doubling time: The amount of time it takes for scales in the universe to double in size with expansion.

Electron volt (eV): A unit of energy equal to the kinetic energy gained by an electron moving under a potential of one volt. The Large Hadron Collider conducts collisions of protons at an energy of ten trillion electron volts.

E-mode polarization: Polarization pattern with no helicity.

Equilibrium: A state of balance in a system, with no dynamics.

Eternal inflation: An inflationary phase dominated by quantum diffusion, for which inflation never ends.

Expansion drag: A dynamical effect in which cosmological expansion acts like a frictional force.

Fine structure constant: The coupling constant in the standard model describing the strength of electromagnetic interactions.

Fine-tuning: The unexplained setting of a fundamental constant to a particular value.

General relativity: Albert Einstein's relativistic theory of gravity.

Ghost instability: An instability of the vacuum associated with the decay into negative energy states.

Gravitational lensing: The bending of light by gravity.

Gravitational waves: Waves propagating in the curvature of space-time.

Ground state: The lowest possible energy state of a quantum system.

Higgs field: The field responsible for the symmetry breaking between the electromagnetic and weak nuclear forces in the standard model of particle physics.

Higgs boson: The particle associated with the Higgs field.

Homogeneity: The property of being the same at all places.

Horizon: The boundary of the observable universe, determined by the cosmological expansion rate.

Horizon problem: The lack of an explanation in standard big bang cosmology for the homogeneity of the universe on scales larger than the horizon size.

Hubble constant: The Hubble parameter in the current universe, usually denoted by the symbol H_0.

Hubble law: The law of cosmic expansion such that the recession velocity is proportional to the distance.

Hubble parameter: A number quantifying the expansion rate of the universe, in units of inverse time.

Inflaton: The field responsible for inflation.

Isotropy: The property of being the same in all directions.

Lambda cold dark matter (LCDM) cosmology: The standard big bang cosmological model, for which the universe consists of about 69 percent dark energy, 26 percent dark matter, and 5 percent baryonic matter.

Landscape (also string landscape): The potential surface associated with the large number of possible compactifications of extra dimensions in string theory.

Lensing: Short for gravitational lensing, or the bending of light by gravity.

Lepton: Electrons and their heavy cousins, the μ and τ particles.

Magnetic monopole: A hypothetical particle with nonzero magnetic charge. Not known to exist.

Measure: A definition of probability for a given system.

Measure problem: The inability to specify well-defined probabilities in infinite sets.

Moduli: Fields in string theory that specify how a space is compactified.

Multifield inflation: Inflation driven by more than one scalar field.

Multiverse: A collection of many separate universes. Predicted by eternal inflation.

Non-Gaussianity: Deviation of primordial perturbations from perfectly Gaussian random statistics.

Null energy condition (NEC): The condition in general relativity that all particles must have nonnegative total energy.

Observable universe: The region of the universe within our cosmic horizon, accessible to observation.

Perturbations: Small deviations from homogeneity.

Plasma: A state of matter in which atoms are entirely ionized, consisting of positively charged nuclei and free electrons.

Potential: The energy associated with the state of a scalar field, or another system involving symmetry breaking. Often plotted as a surface.

Primordial perturbations: The initial deviations from homogeneity in the universe that source structure formation. Generated during inflation.

Ptolemaic system: The geocentric cosmology proposed by the ancient Alexandrian astronomer Ptolemy.

Quantum diffusion: Random evolution of a system due to quantum mechanical uncertainty.

Quantum field theory: A quantum mechanical theory describing relativistic fields. An example is the standard model of particle physics.

Quantum gravity: A theory that unifies quantum mechanics and gravity. No such theory currently exists, but candidates include string theory and loop quantum gravity.

Redshift: An increase in wavelength due to cosmological expansion.

Red spectrum: A non–scale invariant spectrum with more power at long wavelengths.

Relic particle: A species of particle "left over" from the early universe that has mostly ceased to interact with other particles.

Scalar field: A quantum field with zero spin.

Scalar perturbations: Perturbations in the density of the universe.

Scale invariance: The property of having the same amplitude at all wavelengths.

Self-coupling: The coupling constant describing the strength of a particle's interaction with other particles of the same type.

Self-reproduction: The process during eternal inflation by which inflationary expansion generates new space more quickly than patches of the universe exit inflation.

Single-field inflation: Inflation driven by one scalar field.

Singularity: A point at which the physical properties of space-time become infinite and consistent theoretical description breaks down.

Slow roll: Phase during inflation for which expansion drag dominates the dynamics of the inflaton field.

Space-time: Space and time conceived according to relativity as a four-dimensional (three space plus one time) object.

Spectral index (also spectral tilt): A number determining the slope of a power-law function. A scale-invariant spectrum has a spectral index $n = 1$. A red spectrum has $n < 1$, and a blue spectrum has $n > 1$.

Standard model: The currently accepted quantum field theory describing the strong, weak, and electromagnetic forces, containing quarks, leptons, and neutrinos.

Superhorizon: Larger than the horizon size of the universe.

Symmetry breaking: A dynamical process by which a system evolves from a state displaying a symmetry to a state in which that symmetry is no longer realized.

Tensor fraction, or tensor/scalar ratio: The fraction of primordial perturbations that consist of gravitational waves. For example, a tensor fraction of 0.1 means that 10 percent of the primordial perturbations are gravitational waves, and the other 90 percent are perturbations in the density of the universe.

Tensor perturbations: Primordial gravitational waves.

Vacuum: A state in quantum field theory containing no particles.

Vacuum energy: The energy associated with a zero-particle or vacuum state in quantum field theory.

Zero-point mode/zero-point fluctuation: Fluctuations associated with the lowest energy state of a quantum system due to the Heisenberg uncertainty principle.

Further Reading

The reader interested in further reading about the physics of the very early universe can look at "The Self-Reproducing Inflationary Universe" by Andrei Linde, *The Inflationary Universe* by Alan Guth, and *At the Edge of Time: Exploring the Mysteries of Our Universe's First Seconds* by Dan Hooper.[1] For general books on cosmology, Steven Weinberg's *The First Three Minutes* remains a classic and is still well worth reading despite its age.[2] The reader interested in a historical treatment of developments in modern cosmology can turn to *Cosmology's Century: An Inside History of Our Modern Understanding of the Universe* by Nobel Laureate P. J. E. Peebles, who was central to much of that history.[3] A more mathematical treatment of basic cosmology, suitable for undergraduate physics students, is Barbara Ryden's lucid and readable *Introduction to Cosmology*.[4] A lively and entertaining treatment of dark matter, dark energy, and the standard cosmological model is in *The Cosmic Cocktail: Three Parts Dark Matter* by Katherine Freese.[5] Readers interested in string theory and other models for quantum gravity are referred to Brian Greene's excellent books *The Elegant Universe: Superstrings, Hidden Dimensions, and the Quest for the Ultimate Theory* and *The Fabric of the Cosmos: Space, Time, and the Texture of Reality* as well as Jim Baggot's *Quantum Space: Loop Quantum Gravity and the Search for the Structure of Space, Time, and the Universe*.[6] The fascinating subject of the ultimate *end* of the universe is covered in *The End of Everything: (Astrophysically Speaking)* by Katie Mack, and *Until the End of Time: Mind, Matter, and Our Search for Meaning in an Evolving Universe* by Brian Greene.[7] Finally, an essential and blistering critique of the (sometimes sorry) state of modern theoretical physics is in Sabine Hossenfelder's marvelous *Lost in Math: How Beauty Leads Physics Astray*.[8]

This list is of course woefully incomplete, and I apologize to the many worthy authors whose work I have neglected in the interest of brevity.

Notes

Chapter 1

1. *The Garden of Epicurus*, trans. Alfred Allinson (John Lane, 1908), https://archive.org/details/gardenofepicurus00fran.

2. Nicolas Copernicus, *On the Revolutions*, trans. Edward Rosen (Baltimore: Johns Hopkins University Press, 1992).

3. Aleksandr Friedmann, "Über die Krümmung des Raumes," *Zeitschrift für Physik* 10, no. 1 (December 1, 1922): 377–386; A. Georges Lemaître, "A Homogeneous Universe of Constant Mass and Increasing Radius Accounting for the Radial Velocity of Extra-Galactic Nebulæ," *Monthly Notices of the Royal Astronomical* Society 91, no. 5 (March 13, 1931): 483–490; Howard P. Robertson, "Kinematics and World-Structure," *Astrophysical Journal* 82 (1935): 284; Robertson, "Kinematics and World-Structure," 284.

4. G. J. Whitrow, "E. A. Milne and Cosmology," *Quarterly Journal of the Royal Astronomical Society* 37 (1996): 365–367, http://adsabs.harvard.edu/full/1996QJRAS..37..365W.

5. Stephen William Hawking, "The Occurrence of Singularities in Cosmology," *Proceedings of the Royal Society of London A: Mathematical and Physical Sciences* 294, no. 1439 (October 18, 1966): 511–521; Stephen William Hawking, "The Occurrence of Singularities in Cosmology II," *Proceedings of the Royal Society of London A: Mathematical and Physical Sciences* 295, no. 1443 (December 20, 1966): 490–493; Stephen William Hawking and Roger Penrose, "The Singularities of

Gravitational Collapse and Cosmology," *Proceedings of the Royal Society A: Mathematical, Physical and Engineering Sciences* 314, no. 1519 (January 1970): 529–548; S. W. Hawking and G. F. R. Ellis, *The Large Scale Structure of Space-Time* (Cambridge: Cambridge University Press, 1973).

6. N. Aghanim, Yashar Akrami, Mark Ashdown, Jonathan Aumont, Carlo Baccigalupi, Mario Ballardini, A. J. Banday, et al., "Planck 2018 Results: VI. Cosmological Parameters," 2018, http://arxiv.org/abs /1807.06209.

Chapter 2

1. A. Georges Lemaître, "Un Univers homogène de masse constante et de rayon croissant rendant compte de la vitesse radiale des nébuleuses extra-galactiques," *Annales de la Société Scientifique de Bruxelles* 47 (1927): 49–59; Edwin Hubble, "A Relation between Distance and Radial Velocity among Extra-Galactic Nebulae," *Proceedings of the National Academy of Sciences of the United States of America* 15, no. 3 (March 15, 1929): 168–173.

2. Licia Verde, Tommaso Treu, and Adam G. Riess, "Tensions between the Early and the Late Universe," *Nature Astronomy* 3 (2019): 891–895, http://arxiv.org/abs/1907.10625.

3. N. Aghanim, Yashar Akrami, Mark Ashdown, Jonathan Aumont, Carlo Baccigalupi, Mario Ballardini, A. J. Banday, et al., "Planck 2018 Results: VI. Cosmological Parameters," 2018, http://arxiv.org/abs /1807.06209.

4. Cormac O'Raifeartaigh and Simon Mitton, "Interrogating the Legend of Einstein's 'Biggest Blunder.'" *Physics in Perspective* 20, no. 4 (December 26, 2018): 318–341.

5. Lawrence .M. Krauss and Michael S. Turner, "The Cosmological Constant Is Back," *General Relativity and Gravitation* 27, no. 11 (November 1, 1995): 1137–1144.

6. Dragan Huterer and Daniel L. Shafer, "Dark Energy Two Decades After: Observables, Probes, Consistency Tests," *Reports on Progress in Physics* 81, no. 1 (January 1, 2018): 016901.

7. Katherine Freese, *The Cosmic Cocktail: Three Parts Dark Matter* (Princeton, NJ: Princeton University Press, 2014).

8. Aghanim et al., "Planck 2018 Results."

9. Daniel Boyanovsky, "Phase Transitions in the Early and the Present Universe: From the Big Bang to Heavy Ion Collisions" (lecture delivered at the Nato Advanced Study Institute, Phase Transitions in the Early Universe: Theory and Observations, Erice, December 16–17, 2000), http://arxiv.org/abs/hep-ph/0102120.

10. M. R. Drout, Anthony L. Piro, B. J. Shappee, C. D. Kilpatrick, J. D. Simon, Claudia Contreras, D. A. Coulter, et al., "Light Curves of the Neutron Star Merger GW170817/SSS17a: Implications for R-Process Nucleosynthesis," *Science* 358, no. 6370 (December 22, 2017): 1570–1574.

11. Aghanim et al., "Planck 2018 Results."

Chapter 3

1. P. A. Oesch, G. Brammer, P. G. van Dokkum, G. D. Illingworth, R. J. Bouwens, I. Labbé, M. Franx, et al., "A Remarkably Luminous Galaxy at Z = 11.1 Measured with Hubble Space Telescope Grism Spectroscopy," *Astrophysical Journal* 819, no. 2 (March 2016): 129–139.

2. Amo A. Penzias and Robert W. Wilson, "A Measurement of Excess Antenna Temperature at 4080 Mc/s," *Astrophysical Journal* 142 (July 1965): 419–421.

3. Penzias and Wilson, "A Measurement of Excess Antenna Temperature at 4080 Mc/s.""

4. David Burstein, S. M. Faber, and Alan Dressler, "Evidence from the Motions of Galaxies for a Large-Scale, Large-Amplitude Flow toward the Great Attractor," *Astrophysical Journal* 354 (1990): 18–32, http://dx.doi.org/10.1086/168664.

5. Edward L. Wright, "History of the CMB Dipole," 2006, http://www.astro.ucla.edu/~wright/CMB-dipole-history.html; Paul S. Henry, "Isotropy of the 3 K Background," *Nature* 231, no. 5304 (June 1971):

516–518; B. E. Corey and David T. Wilkinson, "A Measurement of the Cosmic Microwave Background Anisotropy at 19 GHz," *Bulletin of the American Astronomical Society* 8 (March 1976): 351; George F. Smoot, Marc V. Gorenstein, and R. A. Muller, "Detection of Anisotropy in the Cosmic Blackbody Radiation," *Physical Review Letters* 39, no. 14 (October 3, 1977): 898–901; Marc V. Gorenstein and George F. Smoot, "Large-Angular-Scale Anisotropy in the Cosmic Background Radiation," *Astrophysical Journal* 244 (March 1, 1981): 361–381, https://www.osti.gov/servlets/purl/893033.

6. G. F. Smoot, C. L. Bennett, A. Kogut, E. L. Wright, J. Aymon, N. W. Boggess, E. S. Cheng, et al., "Structure in the COBE Differential Microwave Radiometer First-Year Maps," *Astrophysical Journal* 396 (September 1992): L1–L5; G. Hinshaw, D. N. Spergel, L. Verde, R. S. Hill, S. S. Meyer, C. Barnes, C. L. Bennett, et al., "First-Year Wilkinson Microwave Anisotropy Probe (WMAP) Observations: The Angular Power Spectrum," *Astrophysical Journal Supplement Series* 148, no. 1 (September 2003): 135–159; P. A. R. Ade, N. Aghanim, C. Armitage-Caplan, M. Arnaud, Mark Ashdown, F. Atrio-Barandela, Jonathan Aumont, et al., "Planck 2013 Results: XV. CMB Power Spectra and Likelihood," *Astronomy and Astrophysics* 571 (November 2014).

7. A. D. Sakharov, "The Initial Stage of an Expanding Universe and the Appearance of a Nonuniform Distribution of Matter," *Soviet Journal of Experimental and Theoretical Physics* 22 (1966): 241; A. D. Miller, Robert Caldwell, Mark J. Devlin, W. B. Dorwart, T. Herbig, M. R. Nolta, Lyman Page, et al., "A Measurement of the Angular Power Spectrum of the Cosmic Microwave Background from [CLC][ITAL]l[/ITAL][/CLC] = 100 to 400," *Astrophysical Journal* 524, no. 1 (October 10, 1999): L1–L4; Andrew E. Lange, P. A. R. Ade, J. J. Bock, J. R. Bond, J. Borrill, A. Boscaleri, K. Coble, et al., "Cosmological Parameters from the First Results of Boomerang," *Physical Review D* 63, no. 4 (January 19, 2001): L1.

8. N. Aghanim, Yashar Akrami, Mark Ashdown, Jonathan Aumont, Carlo Baccigalupi, Mario Ballardini, A. J. Banday, et al., "Planck 2018 Results: VI. Cosmological Parameters," 2018, http://arxiv.org/abs/1807.06209.

Chapter 4

1. Adam G. Riess, Alexei V. Filippenko, Peter Challis, Alejandro Clocchiatti, Alan Diercks, Peter M. Garnavich, Ron Gilliland, et al., "Observational Evidence from Supernovae for an Accelerating Universe and a Cosmological Constant," *Astronomical Journal* 116, no. 3 (September 1998): 1009–1038; Saul Perlmutter, G. Aldering, G. Goldhaber, Richard A. Knop, P. Nugent, P. G. Castro, Susana Deustua, et al., "Measurements of Ω and Λ from 42 High-Redshift Supernovae," *Astrophysics* 517, no. 2 (June 1999): 565–586; N. Aghanim, Yashar Akrami, Mark Ashdown, Jonathan Aumont, Carlo Baccigalupi, Mario Ballardini, A. J. Banday, et al., "Planck 2018 Results: VI. Cosmological Parameters," 2018, http://arxiv.org/abs/1807.06209.

2. Alan H. Guth, "Inflationary Universe: A Possible Solution to the Horizon and Flatness Problems," *Physical Review D* 23, no. 2 (January 1981): 347.

3. Alexei A. Starobinsky, "A New Type of Isotropic Cosmological Models without Singularity," *Physics Letters B* 91, no. 1 (March 1980): 99–102; Katsuhiko Sato, "First-Order Phase Transition of a Vacuum and the Expansion of the Universe," *Monthly Notices of the Royal Astronomical Society* 195 (1981): 467–479, http://dx.doi.org/10.1093 /mnras/195.3.467; Katsuhiko Sato, "Cosmological Baryon-Number Domain Structure and the First Order Phase Transition of a Vacuum," *Physics Letters B* 99, no. 1 (February 1981): 66–70.

4. Georges Aad, Tatevik Abajyan, Brad Abbott, Jalal Abdallah, Samah Abdel Khalek, Ahmed Ali Abdelalim, Ovsat Abdinov, et al., "Observation of a New Particle in the Search for the Standard Model Higgs Boson with the ATLAS Detector at the LHC," *Physics Letters B* 716, no. 1 (September 2012): 1–29; Serguei Chatrchyan, Vardan Khachatryan, Albert M. Sirunyan, Armen Tumasyan, Wolfgang Adam, Ernest Aguilo, Thomas Bergauer, et al., "Observation of a New Boson at a Mass of 125 GeV with the CMS Experiment at the LHC," *Physics Letters B* 716, no. 1 (September 2012): 30–61.

5. Alison Wright, "Nobel Prize 2013: Englert and Higgs," *Nature Physics* 9, no. 11 (October 8, 2013): 692.

6. François Englert, "Robert Brout," *Physics Today* 64, no. 8 (August 2011): 63–64.

7. Andrei D. Linde, "Is the Lee Constant a Cosmological Constant?," *Journal of Experimental and Theoretical Physics Letters* 19 (1974): 183; Joseph Dreitlein, "Broken Symmetry and the Cosmological Constant," *Physical Review Letters* 33, no. 20 (November 11, 1974): 1243–1244.

8. Fedor L. Bezrukov and Mikhail E. Shaposhnikov, "The Standard Model Higgs Boson as the Inflaton," *Physics Letters B* 659, no. 3 (January 24, 2008): 703–706.

9. Guth, "Inflationary Universe," 347.

10. Andrei D. Linde, "A New Inflationary Universe Scenario: A Possible Solution of the Horizon, Flatness, Homogeneity, Isotropy and Primordial Monopole Problems," *Physics Letters B* 108, no. 6 (February 1982): 389–393; Andreas Albrecht and Paul J. Steinhardt, "Cosmology for Grand Unified Theories with Radiatively Induced Symmetry Breaking," *Physical Review Letters* 48, no. 17 (April 26, 1982): 1220–1223.

11. Andrei D. Linde, "Chaotic Inflation," *Physics Letters B* 129, nos. 3–4 (September 22, 1983): 177–181.

12. A. Georges Lemaître, "Un Univers homogène de masse constante et de rayon croissant rendant compte de la vitesse radiale des nébuleuses extra-galactiques," *Annales de la Société Scientifique de Bruxelles* 47 (1927): 49–59.

13. Tom W. B. Kibble, "Topology of Cosmic Domains and Strings," *Journal of Physics A: Mathematical and General* 9, no. 8 (August 19, 1976): 1387–1398.

14. Guth, "Inflationary Universe."

15. John Ellis, Andrei D. Linde, and Dimitri V. Nanopoulos, "Inflation Can Save the Gravitino," *Physics Letters B* 118, nos. 1–3 (December 1982): 59–64.

16. Guth, "Inflationary Universe."

17. Linde, "A New Inflationary Universe Scenario"; Albrecht and Steinhardt, "Cosmology for Grand Unified Theories."

18. Albrecht and Steinhardt, "Cosmology for Grand Unified Theories"; Andreas Albrecht, Paul J. Steinhardt, Michael S. Turner, and Frank Wilczek, "Reheating an Inflationary Universe," *Physical Review Letters* 48, no. 20 (May 17, 1982): 1437–1440.

19. Bruce A. Bassett, Shinji Tsujikawa, and David Wands, "Inflation Dynamics and Reheating," *Reviews of Modern Physics* 78, no. 2 (May 24, 2006): 537–589; Mustafa A. Amin, Mark P. Hertzberg, David I. Kaiser, and Johanna Karouby, "Nonperturbative Dynamics of Reheating after Inflation: A Review," *International Journal of Modern Physics D* 24, no. 1 (January 28, 2015): 1530003.

Chapter 5

1. Prakash Sarkar, Jaswant Yadav, Biswajit Pandey, and Somnath Bharadwaj, "The Scale of Homogeneity of the Galaxy Distribution in SDSS DR6," *Monthly Notices of the Royal Astronomical Society* 399, no. 1 (October 1, 2009): L128–L131.

2. Joseph Silk, "Cosmic Black-Body Radiation and Galaxy Formation," *Astrophysical Journal* 151 (February 1968): 459–471.

3. Isaac Newton, *Philosophiae Naturalis Principia Mathematica* (Project Gutenberg edition, 2009), http://www.gutenberg.org/ebooks/28233.

4. Charles W. Misner, Kip S. Thorne, and John Archibald Wheeler, *Gravitation* (Princeton, NJ: Princeton University Press, 2017).

5. W. Heisenberg, "Über den anschaulichen Inhalt der quantentheoretischen Kinematik und Mechanik," *Zeitschrift für Physik* 43 (1927): 172–198, http://dx.doi.org/10.1007/bf01397280.

6. Hendrik B. G. Casimir, "On the Attraction between Two Perfectly Conducting Plates," in *Modern Kaluza-Klein Theories*, Frontiers in Physics, ed. Thomas Appelquist, Alan Chodos, and Peter G. O. Freund (1948; repr., Boston: Addison-Wesley, 1987), 65:342–344.

7. Michael Bordag, Usman Mohideen, and Vladimir M. Moste-panenko, "New Developments in the Casimir Effect," *Physics Reports* 353, nos. 1–3 (October 2001): 1–205.

8. R. P. Mignani, Vincenzo Testa, D. González Caniulef, R. Taverna, Roberto Turolla, Silvia Zane, and K. Wu, "Evidence for Vacuum Birefringence from the First Optical-Polarimetry Measurement of the Isolated Neutron Star RX J1856.5–3754," *Monthly Notices of the Royal Astronomical Society* 465, no. 1 (February 11, 2017): 492–500; Felix Karbstein, "Probing Vacuum Polarization Effects with High-Intensity Lasers," *Particles* 3, no. 1 (January 19, 2020): 39–61.

9. Stephen W. Hawking, "Particle Creation by Black Holes," *Communications in Mathematical Physics* 43, no. 3 (August 1975): 199–220.

10. Gary W. Gibbons and Stephen W. Hawking, "Cosmological Event Horizons, Thermodynamics, and Particle Creation," *Physical Review D* 15, no. 10 (May 15, 1977): 2738–2751.

11. Stephen W. Hawking, "The Development of Irregularities in a Single Bubble Inflationary Universe," *Physics Letters B* 115, no. 4 (September 1982): 295–297; Alan H. Guth and So-Young Pi, "Fluctuations in the New Inflationary Universe," *Physical Review Letters* 49, no. 15 (October 11, 1982): 1110–1113; Alexei A. Starobinsky, "Dynamics of Phase Transition in the New Inflationary Universe Scenario and Generation of Perturbations," *Physics Letters B* 117, nos. 3–4 (November 1982): 175–178; James M. Bardeen, Paul J. Steinhardt, and Michael S. Turner, "Spontaneous Creation of Almost Scale-Free Density Perturbations in an Inflationary Universe," *Physical Review D* 28, no. 4 (August 15, 1983): 679–693.

12. Nikolay N. Bogoljubov, "On a New Method in the Theory of Superconductivity," *Nuovo Cimento* 7, no. 6 (March 1958): 794–805; William G. Unruh, "Notes on Black-Hole Evaporation," *Physical Review D* 14, no. 4 (August 15, 1976): 870–892.

13. B. P. Abbott, R. Abbott, T. D. Abbott, M. R. Abernathy, F. Acernese, K. Ackley, C. Adams, et al., "GW150914: First Results from the Search for Binary Black Hole Coalescence with Advanced LIGO," *Physical Review D* 93, no. 12 (June 7, 2016): 122003.

Chapter 6

1. G. Hinshaw, D. N. Spergel, L. Verde, R. S. Hill, S. S. Meyer, C. Barnes, C. L. Bennett, et al., "First-Year Wilkinson Microwave Anisotropy Probe (WMAP) Observations: The Angular Power Spectrum," *Astrophysical Journal Supplement Series* 148, no. 1 (September 2003): 135–159; P. A. R. Ade, N. Aghanim, C. Armitage-Caplan, M. Arnaud, Mark Ashdown, F. Atrio-Barandela, Jonathan Aumont, et al., "Planck 2013 Results: XV. CMB Power Spectra and Likelihood," *Astronomy and Astrophysics* 571 (November 29, 2014): A15.

2. Kevork Abazajian, Jennifer K. Adelman-McCarthy, Marcel A. Agüeros, Sahar S. Allam, Scott F. Anderson, James Annis, Neta A. Bahcall, et al., "The First Data Release of the Sloan Digital Sky Survey," *Astrophysical Journal Supplement* 126, no. 4 (October 2003): 2081–2086; Shadab Alam, Franco D. Albareti, Carlos Allende Prieto, F. Anders, Scott F. Anderson, Timothy Anderton, Brett H. Andrews, et al., "The Eleventh and Twelfth Data Releases of the Sloan Digital Sky Survey: Final Data from SDSS-III," *Astrophysical Journal Supplement* 219, no. 1 (July 27, 2015): 1–30

3. N. Aghanim, M. Arnaud, Mark Ashdown, Jonathan Aumont, Carlo Baccigalupi, A. J. Banday, Ramon B. Barreiro, et al., "Planck 2015 Results: XI. CMB Power Spectra, Likelihoods, and Robustness of Parameters," *Astronomy and Astrophysics* 594 (October 20, 2016).

4. John M. Kovac, Erik M. Leitch, Clem Pryke, John E. Carlstrom, N. W. Halverson, and William L. Holzapfel, "Detection of Polarization in the Cosmic Microwave Background Using DASI," *Nature* 420, no. 6917 (December 2002): 772–787; N. Aghanim, Yashar Akrami, Mark Ashdown, Jonathan Aumont, Carlo Baccigalupi, Mario Ballardini, A. J. Banday, et al., "Planck 2018 Results. V. CMB Power Spectra and Likelihoods," *Astronomy and Astrophysics* 641 (September 2020).

5. Daniel Baumann, Mark G. Jackson, Peter .Adshead, Alexandre Amblard, Amjad Ashoorioon, Nicola Bartolo, Rachel Bean, et al., "Probing Inflation with CMB Polarization," *AIP Conference Proceedings* 1141 (2009): 10–120.

6. Aghanim et al., "Planck 2018 Results. V. CMB Power Spectra and Likelihoods"; P. A. R. Ade, Z. Ahmed, R. W. Aikin, K. D. Alexander, D. Barkats, S. J. Benton, C. A. Bischoff, et al., "BICEP2/Keck Array VIII: Measurement of Gravitational Lensing from Large-Scale B-Mode Polarization," *Astrophysical Journal* 833, no. 2 (December 19, 2016): 228; P. A. R. Ade, R. W. Aikin, D. Barkats, S. J. Benton, C. A. Bischoff, J. J. Bock, J. A. Brevik, et al., "Detection of B-Mode Polarization at Degree Angular Scales by BICEP2," *Physical Review Letters* 112, no. 24 (June 19, 2014): 467.

7. Scott Dodelson, William H. Kinney, and Edward W. Kolb, "Cosmic Microwave Background Measurements Can Discriminate among Inflation Models," *Physical Review D* 56, no. 6 (September 15, 1997): 3207–3215.

8. Dodelson, Kinney, and Kolb, "Cosmic Microwave Background Measurements Can Discriminate"; P. A. R. Ade, N. Aghanim, M. Arnaud, F. Arroja, Mark Ashdown, Jonathan Aumont, Carlo Baccigalupi, et al., "Planck 2015 Results: XX. Constraints on Inflation," *Astronomy and Astrophysics* 594 (October 20, 2016): A20; P. A. R. Ade, Z. Ahmed, R. W. Aikin, K. D. Alexander, D. Barkats, S. J. Benton, C. A. Bischoff, et al., "BICEP2/Keck Array V: Measurements of B-mode Polarization at Degree Angular Scales and 150 GHz by the Keck Array," *Astrophysical Journal* 811, no. 2 (September 29, 2015): 126.

9. Alexei A. Starobinsky, "A New Type of Isotropic Cosmological Models without Singularity," *Physics Letters B* 91, no. 1 (March 1980): 99–102.

10. Hideaki Kudoh, Atsushi Taruya, Takashi Hiramatsu, and Yoshiaki Himemoto, "Detecting a Gravitational-Wave Background with Next-Generation Space Interferometers," *Physical Review D* 73, no. 6 (March 6, 2006): 064006; Vincent Corbin and Neil J. Cornish, "Detecting the Cosmic Gravitational Wave Background with the Big Bang Observer," *Classical Quantum Gravity* 23, no. 7 (April 7, 2006): 2435–2446; Seiji Kawamura, Masaki Ando, Takashi Nakamura, Kimio Tsubono, Takahiro Tanaka, Ikkoh Funaki, Naoki Szeto, et al., et al., "The Japanese Space Gravitational Wave Antenna—DECIGO," *Classical Quantum Gravity* 23, no. 8 (April 21, 2006): S125–S131.

11. Juan Maldacena, "Non-Gaussian Features of Primordial Fluctuations in Single Field Inflationary Models," *Journal of High Energy Physics* 2003, no. 5 (May 9, 2003): 013.

12. Yashar Akrami, F. Arroja, Mark Ashdown, Jonathan Aumont, Carlo .Baccigalupi, Mario Ballardini, A. J. Banday, et al., "Planck 2018 Results: IX. Constraints on Primordial Non-Gaussianity," *Astronomy and Astrophysics* 641 (September 2020), http://arxiv.org/abs/1905.05697.

13. Akrami et al., "Planck 2018 Results: IX."

14. Asantha R. Cooray, "21-cm Background Anisotropies Can Discern Primordial Non-Gaussianity," *Physical Review Letters* 97, no. 26 (December 26, 2006): 510; David Mimoun, Mark A. Wieczorek, Leon Alkalai, W. Bruce Banerdt, David Baratoux, Jean-Louis Bougeret, Sylvain Bouley, et al., "Farside Explorer: Unique Science from a Mission to the Farside of the Moon," *Experimental Astronomy* 33, nos. 2–3 (April 27, 2012): 529–585.

15. Mark B. Hoffman and Michael S. Turner, "Kinematic Constraints to the Key Inflationary Observables," *Physical Review D* 64, no. 2 (June 11, 2001): 955; William H. Kinney, "Inflation: Flow, Fixed Points, and Observables to Arbitrary Order in Slow Roll," *Physical Review D* 66, no. 8 (October 18, 2002): 347.

16. Fedor L. Bezrukov and Mikhail Shaposhnikov, "The Standard Model Higgs Boson as the Inflaton," *Physics Letters B* 659, no. 3 (January 24, 2008): 703–706.

Chapter 7

1. Saint Thomas Aquinas, *Summa Theologica, Part I (Prima Pars)* (Project Gutenberg edition, 2006), http://www.gutenberg.org/ebooks/17611.

2. Wikimedia, "Giordano Bruno," Wikimedia Foundation, Inc., 2003, https://en.wikiquote.org/wiki/Giordano_Bruno#On_the_Infinite_Universe_and_Worlds_(1584).

3. Dorothea Waley Singer, *Giordano Bruno; His Life and Thought: With Annotated Translation of His Work "On the Infinite Universe and Worlds"* (Westport, CT: Greenwood Press, 1968).

4. Andrei D. Linde, "Eternal Chaotic Inflation," *Modern Physics Letters A* 1, no. 2 (May 25, 1986): 81–85.

5. J. Richard Gott, "Creation of Open Universes from de Sitter Space," *Nature* 295, no. 5847 (January 1982): 304–307; Martin Bucher, Alfred S. Goldhaber, and Neil Turok, "Open Universe from Inflation," *Physical Review D* 52, no. 6 (September 15, 1995): 3314–3337.

6. Hugh Everett, "'Relative State' Formulation of Quantum Mechanics," *Reviews of Modern Physics* 29, no. 3 (July 1, 1957): 454–462.

7. Alexander S. Goncharov, Andrei D. Linde, and Viatcheslav F. Mukhanov, "The Global Structure of the Inflationary Universe," *International Journal of Modern Physics A* 2, no. 3 (June 25, 1987): 561–591; Alan H. Guth, "Eternal Inflation and Its Implications," *Journal of Physics A: Mathematical and Theoretical* 30, no. 25 (June 22, 2007): 6811–6826.

8. Gabriela Barenboim, Wan-II Park, and William H. Kinney, "Eternal Hilltop Inflation," *Journal of Cosmology and Astroparticle Physics* 2016, no. 5 (May 13, 2016): 030; Alexei A. Starobinsky, "A New Type of Isotropic Cosmological Models without Singularity," *Physics Letters B* 91, no. 1 (March 1980): 99–102; Renata Kallosh, Andrei D. Linde, and Diederik Roest, "Universal Attractor for Inflation at Strong Coupling," *Physical Review Letters* 112, no. 1 (January 7 2014): 532; Renata Kallosh, Andrei D. Linde, and Diederik Roest, "Superconformal Inflationary α-Attractors," *Journal of High Energy Physics* 2013, no. 11 (November 27, 2013): 002.

9. *Srimad-Bhagavatam*, SB 6.16.37, PrabhupadaBooks.com, https://prabhupadabooks.com/sb/6/16/37?d=1.

10. Prateek Agrawal, Georges Obied, Paul J. Steinhardt, and Cumrun Vafa, "On the Cosmological Implications of the String Swampland," *Physics Letters B* 784 (September 2018): 271–276; Sumit K. Garg and Chethan Krishnan, "Bounds on Slow Roll and the de Sitter Swampland," *Journal of High Energy Physics* 2019, no. 11 (November 12,

2019): 022; William H. Kinney, Sunny Vagnozzi, and Luca Visinelli, "The Zoo Plot Meets the Swampland: Mutual (In)consistency of Single-Field Inflation, String Conjectures, and Cosmological Data," *Classical Quantum Gravity* 36, no. 11 (June 6, 2019): 117001; Alek Bedroya, Robert Brandenberger, Marilena Loverde, and Cumrun Vafa, "Trans-Planckian Censorship and Inflationary Cosmology," *Physical Review D* 101, no. 10 (2020): 103502.

11. Cliff P. Burgess, "Lectures on Cosmic Inflation and Its Potential Stringy Realizations," *Classical Quantum Gravity* 24, no. 21 (November 7, 2007): S795–S852.

12. Oskar Klein, "The Atomicity of Electricity as a Quantum Theory Law," *Nature* 118, no. 2971 (October 1926): 516.

13. Michael R. Douglas, "The Statistics of String/M Theory Vacua," *Journal of High Energy Physics* 2003, no. 5 (May 19, 2003): 046.

14. Cliff P. Burgess, Michele Cicoli, and Fernando Quevedo, "String Inflation after Planck 2013," *Journal of Cosmology and Astroparticle Physics* 2013, no. 11 (November 5, 2013): 003.

15. Lee Smolin, *The Life of the Cosmos* (Oxford: Oxford University Press, 1998).

16. Matthew Kleban, "Cosmic Bubble Collisions," *Classical Quantum Gravity* 28, no. 20 (October 21, 2011): 204008; Stephen M. Feeney, Matthew C. Johnson, Daniel J. Mortlock, and Hiranya V. Peiris, "First Observational Tests of Eternal Inflation," *Physical Review Letters* 107, no. 7 (August 8, 2011), https://link.aps.org/doi/10.1103/PhysRevLett.107.071301.

17. Albert M. Sirunyan, Armen Tumasyan, Wolfgang Adam, Federico Ambrogi, Ece Asilar, Thomas Bergauer, Johannes Brandstetter, et al., "Combination of Searches for Higgs Boson Pair Production in Proton-Proton Collisions at $\sqrt{s}$ = 13 TeV," *Physical Review Letters* 122, no. 12 (March 29, 2019), https://link.aps.org/doi/10.1103/PhysRevLett.122.121803; Georges Aad, Brad Abbott, Dale Charles Abbott, Adam Abed Abud, Kira Abeling, Deshan Kavishka Abhayasinghe, Syed Haider Abidi, et al., "Combination of Searches for Higgs Boson Pairs in pp Collisions at $\sqrt{s}$ = 13 TeV with the ATLAS Detector," *Physics*

Letters B 800 (January 2020): 135103; Stefano Di Vita, Christophe Grojean, Giuliano Panico, Marc Riembau, and Thibaud Vantalon, "A Global View on the Higgs Self-Coupling," *Journal of High Energy Physics* 2017, no. 9 (September 18, 2017): 094; Alan J. Barr, Matthew J. Dolan, Christoph Englert, Danilo Enoque Ferreira de Lima, and Michael Spannowsky, "Higgs Self-Coupling Measurements at a 100 TeV Hadron Collider," *Journal of High Energy Physics* 2015, no. 2 (February 3, 2015): 321; "Constraints on the Higgs Boson Self-Coupling from ttH+tH, H to gamma gamma Differential Measurements at the HL-LHC," CERN, Report No.: CMS-PAS-FTR-18–020, November 2018, https://cds.cern.ch/record/2647986/files/FTR-18-020-pas.pdf.

18. Steven Weinberg, "The Cosmological Constant Problem," *Reviews of Modern Physics* 61, no. 1 (January 1, 1989): 1–23.

19. Smolin, *The Life of the Cosmos*.

20. Lisa Randall, *Dark Matter and the Dinosaurs: The Astounding Interconnectedness of the Universe* (New York: HarperCollins, 2016).

21. DoD News Briefing: Secretary Rumsfeld and Gen. Myers, Defense.gov transcript, https://archive.defense.gov/Transcripts/Transcript.aspx?TranscriptID=2636.

22. Brandon Carter, "Large Number Coincidences and the Anthropic Principle in Cosmology," in *Confrontation of Cosmological Theories with Observational Data*, ed. Malcolm S. Longair (Dordrecht: Springer, 1974), 291–298.

23. Eran Palti, "The Swampland: Introduction and Review," *Fortschritte der Physik* 67, no. 6 (June 22, 2019): 1900037.

24. Prateek Agrawal, Georges Obied, Paul J. Steinhardt, and Cumrun Vafa, "On the Cosmological Implications of the String Swampland," *Physics Letters B* 784 (September 2018): 271–276; Georges Obied, Hirosi Ooguri, Lev Spodyneiko, and Cumrun Vafa, "De Sitter Space and the Swampland," July 2018, http://arxiv.org/abs/1806.08362; Hirosi Ooguri, Eran Palti, Gary Shiu, and Cumrun Vafa, "Distance and de Sitter Conjectures on the Swampland," *Physics Letters B* 788 (2019): 180–184, http://dx.doi.org/10.1016/j.physletb.2018.11.018.

25. Hiroki Matsui and Fuminobu Takahashi, "Eternal Inflation and Swampland Conjectures," *Physical Review D* 99, no. 2 (January 30, 2019): 023533; William H. Kinney, "Eternal Inflation and the Refined Swampland Conjecture," *Physical Review Letters* 122, no. 8 (February 27, 2019): 081302; Suddhasattwa Brahma and Sarah Shandera, "Stochastic Eternal Inflation Is in the Swampland," *Journal of High Energy Physics* 2019, no. 11 (November 5, 2019): 1–11; Kinney, Vagnozzi, and Visinelli, "The Zoo Plot Meets the Swampland."

Chapter 8

1. Arvind Borde, Alan H. Guth, and Alexander Vilenkin, "Inflationary Spacetimes Are Incomplete in Past Directions," *Physical Review Letters* 90, no. 15 (April 15, 2003): 151301..

2. Tanmay Vachaspati and Mark Trodden, "Causality and Cosmic Inflation," *Physical Review D* 61, no. 2 (December 16, 1999): 347.

3. Andrei D. Linde, "Chaotic Inflation," *Physics Letters B* 129, nos. 3–4 (September 22, 1983): 177–181; Matthew Kleban and Leonardo Senatore, "Inhomogeneous Anisotropic Cosmology," *Journal of Cosmology and Astroparticle Physics* 2016, no. 10 (October 12, 2016): 022.

4. William E. East, Matthew Kleban, Andrei D. Linde, and Leonardo Senatore, "Beginning Inflation in an Inhomogeneous Universe," *Journal of Cosmology and Astroparticle Physics* 2016, no. 9 (September 6, 2016): 010; Katy Clough, Eugene A. Lim, Brandon S. DiNunno, Willy Fischler, Raphael Flauger, and Sonia Paban, "Robustness of Inflation to Inhomogeneous Initial Conditions," *Journal of Cosmology and Astroparticle Physics* 2017, no. 9 (September 18, 2017): 025; Katy Clough, Raphael Flauger, and Eugene A. Lim, "Robustness of Inflation to Large Tensor Perturbations," *Journal of Cosmology and Astroparticle Physics* 2018, no. 5 (May 29, 2018): 065; Josu C. Aurrekoetxea, Katy Clough, Raphael Flauger, and Eugene A. Lim, "The Effects of Potential Shape on Inhomogeneous Inflation," *Journal of Cosmology and Astroparticle Physics* 2020, no. 5 (May 15, 2020): 030.

5. Jerome Martin and Robert H. Brandenberger, "Trans-Planckian Problem of Inflationary Cosmology," *Physical Review D* 63, no. 12

(May 2, 2001): 347; Robert H. Brandenberger and Jerome Martin, "The Robustness of Inflation to Changes in Super-Planck-Scale Physics," *Modern Physics Letters A* 16, no. 15 (May 20, 2001): 999–1006; Jens C. Niemeyer, "Inflation with a Planck-Scale Frequency Cutoff," *Physical Review D* 63, no. 12 (May 2, 2001): 5112.

6. Alek Bedroya and Cumrun Vafa, "Trans-Planckian Censorship and the Swampland," *Journal of High Energy Physics* 2020, no. 9 (September 18, 2020), http://link.springer.com/10.1007/JHEP09(2020)123.

7. William G. Unruh, "Sonic Analogue of Black Holes and the Effects of High Frequencies on Black Hole Evaporation," *Physical Review D* 51, no. 6 (March 15, 1995): 2827–2838.

8. Lam Hui and William H. Kinney, "Short Distance Physics and the Consistency Relation for Scalar and Tensor Fluctuations in the Inflationary Universe," *Physical Review D* 65, no. 10 (April 29, 2002): 347; Richard Easther, Brian R. Greene, William H. Kinney, and Gary Shiu, "Generic Estimate of Trans-Planckian Modifications to the Primordial Power Spectrum in Inflation," *Physical Review D* 66, no. 2 (July 22, 2002): 999; Gia Dvali, Antonio Kehagias, and Antonio Riotto, "Inflation and Decoupling," 2020, http://arxiv.org/abs/2005.05146.

9. Alexei A. Starobinsky, "Robustness of the Inflationary Perturbation Spectrum to Trans-Planckian Physics," *Journal of Experimental and Theoretical Physics Letters* 73, no. 8 (April 2001): 371–374.

10. Ghazal Geshnizjani, William H. Kinney, and Azadeh Moradinezhad Dizgah, "General Conditions for Scale-Invariant Perturbations in an Expanding Universe," *Journal of Cosmology and Astroparticle Physics* 2011, no. 11 (November 30, 2011): 049.

11. David Wands, "Duality Invariance of Cosmological Perturbation Spectra," *Physical Review D* 60, no. 2 (June 11, 1999): 682.

12. Aaron M. Levy, "Fine-Tuning Challenges for the Matter Bounce Scenario," *Physical Review D* 95, no. 2 (January 2017), https://link.aps.org/doi/10.1103/PhysRevD.95.023522.

13. Justin Khoury, Burt A. Ovrut, Paul J. Steinhardt, and Neil Turok, "Ekpyrotic Universe: Colliding Branes and the Origin of the Hot Big Bang," *Physical Review D* 64, no. 12 (November 28, 2001): 347.

14. Anna Ijjas, Jean-Luc Lehners, and Paul J. Steinhardt, "General Mechanism for Producing Scale-Invariant Perturbations and Small Non-Gaussianity in Ekpyrotic Models," *Physical Review D* 89, no. 12 (June 25, 2014), https://link.aps.org/doi/10.1103/PhysRevD.89 .123520.

15. Anna Ijjas and Paul J. Steinhardt, "A New Kind of Cyclic Universe," *Physics Letters B* 795 (August 2019): 666–672.

16. Roger Penrose, "On the Gravitization of Quantum Mechanics 2: Conformal Cyclic Cosmology," *Foundations of Physics* 44, no. 8 (August 27, 2014): 873–890.

17. Richard C. Tolman, "On the Theoretical Requirements for a Periodic Behaviour of the Universe," *Physical Review* 38, no. 9 (November 1, 1931): 1758–1771.

18. Anthony Aguirre and Steven Gratton, "Steady-State Eternal Inflation," *Physical Review D* 65, no. 8 (March 29, 2002): 252; Anthony Aguirre and Steven Gratton, "Inflation without a Beginning: A Null Boundary Proposal," *Physical Review D* 67, no. 8 (April 29, 2003): 2504.

19. From the Turing Digital Archive, at http://www.turingarchive .org/viewer/?id=25&title=1e.

20. Giordano Bruno, *Cause, Principle, and Unity: Five Dialogues*, trans. Robert de Lucca (Cambridge: Cambridge University Press, 1998).

21. Lao Tzu, *Tao Te Ching* (New York: Random House, 2011), 2.

Further Reading

1. Andrei D. Linde, "The Self-Reproducing Inflationary Universe," *Scientific American* 271, no. 5 (November 1994): 48–55; Alan H. Guth, *The Inflationary Universe* (New York: Basic Books, 1998); Dan Hooper, *At the Edge of Time: Exploring the Mysteries of Our Universe's First Seconds* (Princeton, NJ: Princeton University Press, 2019).

2. Steven Weinberg, *The First Three Minutes: A Modern View of the Origin of the Universe* (New York: Basic Books, 1993).

3. P. J. E. Peebles, *Cosmology's Century: An Inside History of Our Modern Understanding of the Universe* (Princeton, NJ: Princeton University Press, 2020).

4. Barbara Ryden, *Introduction to Cosmology* (Cambridge: Cambridge University Press, 2017).

5. Katherine Freese, *The Cosmic Cocktail: Three Parts Dark Matter* (Princeton, NJ: Princeton University Press, 2014).

6. Brian Greene, *The Elegant Universe: Superstrings, Hidden Dimensions, and the Quest for the Ultimate Theory* (New York: W. W. Norton and Company, 2010); Brian Greene, *The Fabric of the Cosmos: Space, Time, and the Texture of Reality* (New York: Knopf, 2004); Jim Baggott, *Quantum Space: Loop Quantum Gravity and the Search for the Structure of Space, Time, and the Universe* (Oxford: Oxford University Press, 2018).

7. Katie Mack, *The End of Everything: (Astrophysically Speaking)* (New York: Simon and Schuster, 2020); Brian Greene, *Until the End of Time: Mind, Matter, and Our Search for Meaning in an Evolving Universe* (New York: Random House, 2020).

8. Sabine Hossenfelder, *Lost in Math: How Beauty Leads Physics Astray* (New York: Basic Books, 2020).

Index